ADVANCES IN PERVASIVE COMPUTING AND NETWORKING

ADVANCES IN PERVASIVE COMPUTING AND NETWORKING

Edited by

BOLESLAW K. SZYMANSKI

Rensselaer Polytechnic Institute

BÜLENT YENER

Rensselaer Polytechnic Institute

Editors:

Boleslaw K. Szymanski
Bülent Yener
Rensselaer Polytechnic Institute
Department of Computer Science
Troy, NY USA

Advances in pervasive computing and networking / edited by Boleslaw K. Szymanski, Bülent Yener.
p. cm.
Includes bibliographical references and index.
ISBN 1-4899-9514-5
1. Ubiquitous computing. 2. Mobile computing. 3. Sensor networks. I. Szymanski, Boleslaw. II. Yener, Bülent, 1959-

QA76.5915.A38 2004
004--dc22

2004059513

ISBN 1-4899-9514-5 ISBN 0-387-23466-7 (eBook) Printed on acid-free paper.

9 8 7 6 5 4 3 2 1 SPIN 11054641

springeronline.com

Contents

List of Figures

List of Tables

Contributing Authors

Jinane Abounadi is a Post-doctoral Lecturer in the Laboratory for Information and Decision Systems, Massachusetts Institute of Technology.

Murat Alanyali is an Assistant Professor in the Department of Electrical and Computer Engineering, Boston University.

Kenneth P. Birman is a Professor in the Department of Computer Science, Cornell University.

Joel Branch is a Ph.D. student in the Computer Science Department, Rensselaer Polytechnic Institute.

Iacopo Carreras is a Researcher in the CreateNet Research Consortium, Trento, Italy.

Gilbert Chen is a Post-doctoral Researcher in the Center for Pervasive Computing and Networking, Rensselaer Polytechnic Institute.

Ling Jyh Chen is a Ph.D. student in the Department Computer Science, University of California at Los Angeles.

Imrich Chlamtac is the President of the CreateNet Research Consortium and the Distinguished Chair in Telecommunications of the Erik Jonsson School of Engineering and Computer Science, The University of Texas at Dallas and Honorary Bruno Kessler Professor with the University of Trento.

Alan Demers is a Professor in the Department of Computer Science, Cornell University.

Johannes Gehrke is an Assistant Professor in the Department of Computer Science, Cornell University.

Mario Gerla is a Professor in the Department of Computer Science, University of California at Los Angeles.

Saikat Guha is a Ph.D. student in the Department of Computer Science, Cornell University.

Amir E. Khandani is a graduate student in the Laboratory for Information and Decision Systems, Massachusetts Institute of Technology.

Bhaskar Krishnamachari is an Assistant Professor in the Electrical Engineering Department, University of Southern California.

Xiaojun Lin is a Ph.D. student in the School of Electrical and Computer Engineering, Purdue University.

Eytan Modiano is an Associate Professor in the Department of Aeronautical and Astronautics as well as in the Laboratory for Information and Decision Systems, Massachusetts Institute of Technology.

Chilukuri K. Mohan is a Professor in the Department of Electrical Engineering and Computer Science, Syracuse University.

Rohan Murty is an undergraduate student in the Department of Computer Science, Cornell University.

Michael J. Pflug is a graduate student in the Computer Science Department, Rensselaer Polytechnic Institute.

Rajmohan Rajaraman is an Associate Professor in the College of Computer and Information Science, Northeastern University.

Sumit Roy is a Professor in the Department of Electrical Engineering, University of Washington.

Venkatesh Saligrama is an Assistant Professor in the Department of Electrical and Computer Engineering, Boston University.

Andreas Savvides is an Assistant Professor in the Electrical Engineering and Computer Science Departments, Yale University.

Sergio D. Servetto is an Assistant Professor in the School of Electrical and Computer Engineering, Cornell University.

Baharak Shademan was an M.S. student in the Electrical Engineering Department, University of Southern California, when preparing the chapter.

Ness B. Shroff is a Professor in the School of Electrical and Computer Engineering, Purdue University.

Mani Srivastava is a Professor in the Electrical Engineering Department, University of California at Los Angeles.

Tony Sun is a Ph.D. student in the Department of Computer Science, University of California at Los Angeles.

Boleslaw K. Szymanski is the Director of the Center for Pervasive Computing and Networking and a Professor in the Computer Science Department, Rensselaer Polytechnic Institute.

Niki Trigoni is a Post-doctoral Associate in the Department of Computer Science, Cornell University.

Lav R. Varshney is an undergraduate student in the School of Electrical and Computer Engineering, Cornell University.

Pramod K. Varshney is the Research Director of the New York State Center for Advanced Technology in Computer Applications and Software Engineering and a Professor in the Department of Electrical Engineering and Computer Science, Syracuse University.

Hagen Woesner is a Researcher in the CreateNet Research Consortium, Trento, Italy.

Guang Yang is a Ph.D. student in the Department of Computer Science, University of California at Los Angeles.

Yong Yao is a Ph.D. student in the Department of Computer Science, Cornell University.

Lizhong Zheng is an Assistant Professor in the Department of Electrical Engineering and Computer Science, Massachusetts Institute of Technology.

Congzhou Zhou was an M.S. student in the Electrical Engineering Department, University of Southern California, while preparing the chapter.

Jing Zhu is a Senior Research Scientist at the Communications Technology Laboratory, Intel Corporation.

Lijuan Zhu is a Ph.D. student in the Computer Science Department, Rensselaer Polytechnic Institute.

Preface

Pervasive Computing and Networking aim at providing ubiquitous, ever-present, adaptable and smart services enveloping immerse interactions of computing components and/or humans over over large spaces. The computing might of processors is combined with the communication power of the wireless and mobile networking yielding novel environments and paradigms. To fully realize the potential of such environments, many technical and economical challenges needs to be overcome. These challenges and the perspective on the seminal directions of the research in this area were the leading subjects of the Workshop for Pervasive Computing and Networking, sponsored by the National Science Foundation and organized by the Center for Pervasive Computing and Networking in Rensselaer Polytechnic Institute, Troy, New York, USA. This book includes thirteen chapters based on the presentations made at the workshop by the leaders in the field. The scope of the topics ranges from the fundamental theory of the pervasive computing and networking to the crucial challenges of the emerging applications.

The first four chapters of the book address the fundamental challenges of the field. Chapter 1 proposes a scalable, self-organizing technology for sensor networks. The authors present an approach that enables sensor networks to organize themselves automatically and to adaptively counter changing environmental conditions, failures, intermittent connectivity, and limitation of power supply. Chapter 2 discusses the fundamental relationship between the achievable capacity and delay in mobile wireless networks. First, the authors establish an upper bound on capacity achievable under the given delay. Then they show that this bound is tight by designing a scheduling scheme that can achieve it up to a logarithmic factor. In Chapter 3, the authors advocate the use of self-optimization in sensor networks. They demonstrate on several examples that sensor networks can autonomously learn application-specific information through sensor and network observations during the course of their operation. The knowledge gained in this learning can then be then used to self-optimize system performance over time. Finally, Chapter 4 focuses on biological kinetic service oriented networks. It explores the similarities and differences between computer networks and their biological counterparts. The authors argue that both novel approaches to network element design and their behavioral rules can be inspired by the biological networks that were perfected over millions of years of evolution.

The next group of chapters focus on communication challenges in mobile, wireless and sensor networks. Chapter 5 is devoted to cooperative routing in wireless networks. It discusses this problem jointly with transmission-side diversity. The authors study the benefits of coordinating transmission between multiple nodes equipped with a single omni-directional antenna. Chapter 6 investigates the use of in-network data processing for classification of an unknown event using noisy sensor measurements. The authors describe in-network decision making algorithm based on local message-passing that minimizes the communication. Chapter 7 describes methods for maximizing throughput of 802.11 based mesh networks. For this purpose, the authors employ tuning of physical carrier sensing and propose to exploit multiple radios per node technology as it becomes available. In Chapter 8, the authors address the issue of self-configuring location discovery in smart environments. Authors' approach is based on their experiences with the design and use of a rapidly installable self-configuring beaconing system. Finally, Chapter 9 proposes a new direction in multi-query optimization for sensor networks based on a powerful new database abstraction in which clients can "program" the sensors through queries in a high-level declarative language.

Subsequent chapters address three major applications of sensor networks: video streaming, smart buildings, and utility infrastructures. Chapter 10 focuses on ubiquitous video streaming. The authors study adaptive video streaming in vertical handoff scenarios and demonstrate that it results in improvement of user-perceived video quality. Another important application of sensor networks, addressed in Chapter 11, is the monitoring and control of indoor environments, with objectives such as occupant comfort maximization and health risk minimization. Chapter 12 describes a design for a distributed transmitter for reachback based on radar signals sensing and two-radio multi-channel clustering. The authors evaluate the performance of their solution using both information theoretic and simulation-based models.

The book closes with description of a new sensor network simulator in Chapter 13. The authors use a novel component-based simulation guided by the component-port model and simulation component classifications. The developed simulator, called SENSE, provides a new simulation tool that is both easy to use and efficient to execute.

Together, the chapters present a review of the current issues, challenges and directions in the newly emerged area of pervasive computing and networking.

BOLESLAW K. SZYMANSKI AND BÜLENT YENER

Acknowledgments

The chapters of this book were based on the presentations at the NSF-RPI Workshop on Pervasive Computing and Networking held in Troy, New York in April 2004. The editors and the authors of the chapters express their gratitude to the National Science Foundation for the financial support provided for the Workshop through the NSF Grant CNS-0340877. The content of the papers is sole responsibility of their authors, and these papers do not necessarily reflect the position or policy of the U.S. Government—no official endorsement should be inferred or implied.

Chapter 1

SCALABLE, SELF-ORGANIZING TECHNOLOGY FOR SENSOR NETWORKS

Kenneth P. Birman, Saikat Guha and Rohan Murty
Department of Computer Science
Cornell University, Ithaca, NY 14853
{ken,sg266,rnm5}@cs.cornell.edu

Abstract Sensor networks will often need to organize themselves automatically and adapt to changing environmental conditions, failures, intermittent connectivity, and in response to power considerations. We review a series of technologies that we find interesting both because they solve fundamental problems seen in these settings, and also because they appear to be instances of a broader class of solutions responsive to these objectives.

Keywords: Embedded computing, sensors, distributed monitoring, routing, network position information.

1. Introduction

The emergence of a new generation of technologies for pervasive computing and networking is challenging basic assumptions about how networked applications should behave. Wired systems typically ignore power and location considerations and operate "in the dark" with respect to overall system configuration, current operating modes or detected environmental properties, and positions of devices both in absolute and logical terms. These kinds of assumptions represent serious constraints and lead to sub-optimal solutions in embedded or pervasive computing systems.

At Cornell, we and other researchers are working to develop platform technologies responsive to these and related considerations. This paper reports on three representative examples, which we offer with two goals in mind. First, each of these technologies reflects a mixture of properties and algorithmic features matching the special requirements seen in pervasive computing settings. Second, we believe that the underlying methodologies reflected in the three technologies are interesting in themselves, because they point to broader opportunities for future study.

At the time of this writing, the technologies are not integrated into a single platform, but doing so is an eventual goal. Indeed, we believe that over time, researchers will conclude that pervasive computing systems demand a completely new kind of infrastructure, built using components such as the ones we present here.

Specific properties of importance include the following. First, our solutions are self-organizing, a crucial property in many emerging applications. They are strongly self-stabilizing, converging rapidly to a desired structure and repairing themselves rapidly after disruption. They lend themselves to theoretical modeling and analysis, lending themselves to both pencil-and-paper study and to simulation. Significantly (and unlike many distributed systems technologies), the analyses so obtained hold up well in practice; as we'll see below, this is because our protocols are so overwhelmingly convergent. Moreover, they are robust to perturbation, a property that may be extremely important in the relatively turbulent world in which many sensor applications will need to operate. Interestingly, each solution consists of a relatively simple protocol run in parallel by the components of the system, and the desired global outcome "emerges" rapidly through the interaction of a component with its neighbors. We conjecture that a rich class of solutions having these properties awaits discovery by future researchers.

The three services on which we focus here are (1) *Astrolabe*, a system for distributed state monitoring, application management, and data mining constructed using a novel peer-to-peer protocol that offers unique scalability, low load, and rapid convergence; (2) *Tycho,* a location-aware event localization system for sensor networks, and (3) *Sextant,* a system for discovering sensor locations using software methods that is highly accurate, energy-efficient and scalable. Each is really an instance from a broader class of related solutions, and is interesting both in its own terms, but also as exemplars of these broader classes.

2. Astrolabe

The Astrolabe system is best understood as a relational database built using a peer-to-peer protocol running between the applications or computers on which Astrolabe is installed. Like any relational database, the fundamental building block employed by Astrolabe is a tuple (a row of data items) into which values can be stored. For simplicity in this paper, we'll focus on the case where each tuple contains information associated with some computer. The technology is quite general, however, and can be configured with a tuple per application, or even with a tuple for each instance of some type of file or database. For the purposes of this paper, Astrolabe would be used to capture and track information associated with sensors in a network of sensors. For reasons of brevity, this section (and those that follow) limits itself to a brief overview; additional detail can be found in [1, 2].

The data stored into Astrolabe can be drawn from any of a number of sources. Small sensors would export data "directly" but a larger, more comprehensive computing node could export more or less any kind of data that can be encoded efficiently. This includes information in the management information base (MIB), fields extracted directly from a file, database, spreadsheet, or information fetched from a user-supplied method associated with some application program. Astrolabe is also flexible about data types, supporting the usual basic types but also allowing the application to supply arbitrary information encoded with XML. The only requirement is that the total size of the tuple be no more than a few k-bytes; much larger objects should be handled outside the core Astrolabe framework.

The specific data that should be pulled into Astrolabe is specified in a configuration certificate. Should the needs of the user change, the configuration certificate can be modified and, within a few seconds, Astrolabe will reconfigure itself accordingly. This action is, however, restricted by a security policy, details of which are described in [1, 2].

Astrolabe groups small sets of tuples into a hierarchy of relational tables. A "leaf" table consists of perhaps 30 to 60 tuples (we could scale up to hundreds but not thousands of types in a single table) containing data from sources physically close to one-another in the network. This grouping (a database administrator would recognize it as a form of schema) can often be created automatically, using latency and network addresses to identify nearby machines (the location information could, for example, be obtained using the method we present in Section 4 of this paper).

The data collected by Astrolabe evolves as the underlying information sources report updates, hence the system constructs a continuously changing database using information that actually resides on the participating computers. Figure 1.1 illustrates this: we see a collection of small database relations, each tuple corresponding to one machine, and each relation collecting tuples associated with some set of nearby machines. In this figure, the data stored within the tuple includes the name of the machine, its current load, an indication of whether or not various servers are running on it, and the "version" for some application. Keep in mind that this selection of data is completely determined by the configuration certificate. In principle, any data available on the machine or in any

Astrolabe builds a hierarchy using a P2P protocol that "assembles the puzzle" without any servers

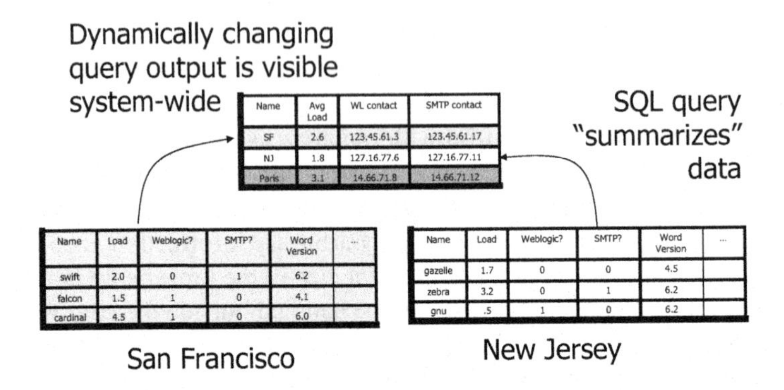

Name	Avg Load	WL contact	SMTP contact
SF	2.6	123.45.61.3	123.45.61.17
NJ	1.8	127.16.77.6	127.16.77.11
Paris	3.1	14.66.71.8	14.66.71.12

Name	Load	Weblogic?	SMTP?	Word Version	...
swift	2.0	0	1	6.2	
falcon	1.5	1	0	4.1	
cardinal	4.5	1	0	6.0	

Name	Load	Weblogic?	SMTP?	Word Version	...
gazelle	1.7	0	0	4.5	
zebra	3.2	0	1	6.2	
gnu	.5	1	0	6.2	

Figure 1.1. (a) Query Area (b) Subgraph (c) Tree

application running on the machine can be exported. In particular, spreadsheets and databases can easily be configured to export data to Astrolabe.

The same interfaces which enable us to fetch data so easily also make it easy for applications to use Astrolabe. Most commonly, an application would access the Astrolabe relations just as it might access any other table, database or spreadsheet. As updates occur, the application receives a form of event notifying it that the table should be rescanned. Thus, with little or no specialized programming, data from Astrolabe data could be " dragged " into a local database, spreadsheet, or even onto a web page. As the data changes, the associated application will receive refresh events.

Astrolabe is intended for use in very large networks, hence this form of direct access to local data cannot be used for the full dataset: while the system does capture data throughout the network, the amount of information would be unwieldy and the frequency of updates excessive. Accordingly, although Astrolabe does provide an interface whereby a remote region's data can be accessed, the normal way of monitoring remote data is through aggregation queries.

An aggregation query is, as the name suggests, just an SQL query which operates on these leaf relations, extracting a single summary tuple from each which reflects the globally significant information within the region. Sets of summary tuples are concatenated by Astrolabe to form summary relations (again, the size is typically 30 to 60 tuples each), and if the size of the system is large enough so that there will be several summary relations, this process is repeated at the next level up, and so forth. Astrolabe is thus a hierarchical relational database, and this is also visible in Figure 1.1, where the summaries of the various regions appear as rows in the root relation. Each of the summaries is updated, in real-time, as the leaf data from which it was formed changes. Even in networks with thousands or millions of instrumented machines, updates are visible system-wide within a few tens of seconds. Since sensor networks may be very large, this scalability is likely to be important.

A computer using Astrolabe will, in general, keep a local copy of the data for its own region and aggregation (summary) data for region above it on the path to the root of this hierarchy. As just explained, the system maintains the abstraction of a hierarchical relational database. Physically, however, this hierarchy is an illusion, constructed using a peer-to-peer protocol, somewhat like a jig-saw puzzle in which each computer has ownership of one piece and read-only replicas of a few others. Our protocols permit the system to assemble the puzzle as a whole when needed. Thus, while the user thinks of Astrolabe as a somewhat constrained but rather general database, accessed using conventional programmer APIs and development tools, this abstraction is actually an illusion, created on the fly. In particular, the memory needed to run the system is very small, even in a network that may be very large.

The peer-to-peer protocol used for this purpose is, to first approximation, easily described. Each Astrolabe system keeps track of the other machines in its zone, and of a subset of contact machines in other zones. This subset is selected in a pseudo-random manner from the full membership of the system (again, a peer-to-peer mechanism is used to track approximate membership ; for simplicity of exposition we omit any details here). At some fixed frequency, typically every 2 to 5 seconds, each participating machine sends a concise state description to a randomly selected destination within this set of neighbors and remote contacts. The state description is very compact and lists versions of objects available from the sender. We call such a message a " gossip " event. Unless an object is very small, the gossip event will not contain the data associated with it.

Upon receiving a gossip message, an Astrolabe system is in a position to identify information which may be stale at the sender's machine (because timestamps are out of date) or that may be more current at the sender than on its own system. We say may because time elapses while messages traverse the network, hence no machine actually has current information about any other. Our protocols are purely asynchronous: when sending a message, the sender does not pause to wait for it to be received and, indeed, the protocol makes no effort to ensure that gossip gets to its destinations.

If a receiver of a gossip message discovers that it has data missing at the sender machine, a copy of that data is sent back to the sender. We call this a push event. Conversely, if the sender has data lacking at the receiver, a pull event occurs: a message is sent requesting a copy of the data in question. Again, these actions are entirely asynchronous; the idea is that they will usually be successful, but if not (e.g. if a message is lost in the network, received very late, or if some other kind of failure occurs), the same information will probably be obtained from some other source later.

One can see that through exchanges of gossip messages and data, information should propagate within a network over an exponentially increasing number of randomly selected paths among the participants. That is, if a machine updates its own row, after one round of gossip, the update will probably be found at two machines. After two rounds, the update will probably be at four machines, etc. In general, updates propagate in log of the system size – seconds or tens of seconds in our implementation. In practice, we configure Astrolabe to gossip rapidly within each zone (to take advantage of the presumably low latency) and less frequently between zones (to avoid overloading bottlenecks such as firewalls or shared network links). The effect of these steps is to ensure that the communication load on each machine using Astrolabe and also each communication link involved is bounded and independent of network size.

We've said that Astrolabe gossips about objects. In our work, a tuple is an object, but because of the hierarchy used by Astrolabe, a tuple would only be of interest to a receiver in the same region as the sender. In general, Astrolabe gossips about information of shared interest to the sender and receiver. This could include tuples in the regional database, but also aggregation results for aggregation zones that are ancestors of both the sender and receiver.

After a round of gossip or an update to its own tuple, Astrolabe recomputes any aggregation queries affected by the update. It then informs any local readers of the Astrolabe objects in question that their values have changed, and the associated application rereads the object and refreshes its state accordingly. The change would be expected to reach the server within a delay logarithmic in the size of the network, and proportional to the gossip rate. Using a 2-second gossip rate, an update would thus reach all members in a system of 10,000 computers in roughly 25 seconds. Of course, the gossip rate can be tuned to make the system run faster, or slower, depending on the importance of rapid responses and the available bandwidth.

Astrolabe was originally developed for use in very large-scale wired environments, but has several features well-matched to sensor networks and other embedded applications. First, most communication occurs between a components and nearby peers. In wireless ad-hoc routed networks, this is important because sending a message to a very remote component consumes power not just on the sender and receiver, but also on intermediary nodes involved in routing the packet. In a reasonably dense sensor network, most Astrolabe communication will occur between sensors near to one-another, with only aggregation information being transmitted over long distances.

Astrolabe doesn't rely on any single or even "primary" route between components that share information through its database; instead, within any zone data travels over all possible paths within that zone. This is important because it makes the protocol extremely robust to routing disturbances or transient communication problems. Astrolabe will report events within roughly the same amount of time even if serious disruption is occurring and the system repairs itself rapidly after failure or other stresses.

Finally, Astrolabe can be made to configure itself entirely automatically, using proximity within the network to define zones. In the sections that follow we'll see other uses of location information as an input to system configuration algorithms; we believe the idea is one that merits further study and broader use.

Astrolabe is just one of several technologies we've constructed using this methodology. Others relevant to pervasive computing include Bimodal Multicast [3], a scalable protocol that uses peer-to-peer epidemic protocols to achieve very high reliability at rather low cost, and Kelips [4], a novel distributed indexing mechanism (a "DHT" in the current peer-to-peer vocabulary). Kelips can find information for the cost of a single RPC even in a massive network. All of these solutions share strong similarities: inexpensive gossip-based protocols that converge because they mimic the propagation of an epidemic, constant background overheads (on component nodes and links), a preference for local communication, and very robust behavior even under stress. Moreover, precisely because of their overwhelmingly rapid convergence, even simplified theoretical models and analysis tend to be quite robust, yielding predictions that are later confirmed experimentally. Finally, all of these mechanisms have very simple implementations, small memory footprints, and use relatively low bandwidth.

Work still remains to be done: none of our protocols is able to deal with scheduled sleep periods or other power conservation and scheduling considerations. Nonetheless, we believe that they represent exciting starting points.

3. Interactions Between Power Aware Components

A typical sensor tends to be very small in size. Though this serves as an advantage, it also limits the sensor's capabilities in terms of energy and data storage, processing power, and communication capabilities. As sensors mature, we believe that they will become faster, cheaper and be able to collect and store more data and transmit it farther; however, their energy capacity will not increase at a similar rate. There have been advances in the use of wireless power, and renewable sources of energy in sensors, but the technologies are yet to emerge from their infancy. In such light we believe that application and software level power conservation can help balance the energy budget enabling sensor networks to last longer than previously imagined.

We outline below some characteristics of power aware components for sensor-networks that aim to lengthen network lifetime.

- **Hardware re-use**: When possible, existing hardware should be used by multiple components instead of power-consuming single-purpose hardware.
- **Uniform energy dissipation**: Computational and communication load of a node should be proportional to the energy available to it.
- **Low duty-cycle**: Components should periodically allow some nodes to operate in power-saving mode.
- **Reactive protocols**: Reactive protocols or proactive protocols with localized effect should be used to minimize communication costs.
- **Configurable trade-off**: Network administrators should be able to trade-off system lifetime with system latency and effectiveness.

Research has shown communication to be far more expensive than computation for sensor networks [5]. Additionally, unreliable radio communication links further aggravate this problem. In Tycho [9], an event localization system for sensor-networks, we explore the case of a power-aware component where nodes send data to a single controller. In Tycho, a query is injected into the network and upon detection of an event, sensor nodes send the detected information to the controller. Therefore it is desirable to optimize communication links from multiple sources of data to a single destination (controller).

In the operation of a typical sensor network, various components such as the application, routing, location discovery etc. interact with one another. Research in this area has focused on minimizing energy consumption of each of these components when executing independently and when interacting with one another. In this section, we focus on minimizing the energy consumption through increased interactions between the application and the routing layers.

Consider a large sensor network in which nodes can either detect an event or be queried based on certain parameters. A simple approach to reporting results when an event occurs or when a query is passed on to the network, is to flood the controller with messages from each individual sensor node. While this is a simple enough mechanism to code into sensors, it results in inefficient usage of energy, and can fail to report accurate results when a sensor is not capable of communicating directly with the controller. Further, this approach does not scale well with an increase in the number of nodes in the network.

As a result, significant research effort in this area aims to improve the longevity of sensor nodes by optimizing for the sensor's power usage, minimizing communication and the size of messages transferred, without compromising on the functionality of the sensors. This is commonly achieved by installing a routing fabric in the network and then aggregating data along the routing path. An optimal routing protocol tends to either minimize the number of nodes involved in routing or minimize the distance each message is transmitted between adjacent nodes in the network. Energy can be further conserved by aggregating along the path to the controller. This approach is optimal in terms of the total energy consumed by the various nodes to either transmit or receive messages. Further, aggregation has the potential to reduce the number of messages transmitted between nodes as well as the size of the messages.

Various simulations and physical experiments have shown previously proposed protocols to have extended the life of a sensor network well beyond that of direct communication or the flooding approach previously described.

Routing protocols tend to expose functionality through limited interactions with the application. The first of such functionality is the decision regarding which nodes can be used in low duty cycle modes. A large fraction of energy savings are achieved by enabling a significant fraction of the nodes in the network to shift to a low duty cycle mode. While various routing protocols achieve low duty cycle nodes by putting various nodes in the network to sleep, other routing protocols do not achieve low duty cycle nodes since they require all nodes in the network to be alive during routing. A common scheme among routing protocols that support low duty cycle nodes is to interact with the application when deciding on the subset of nodes that are to go to sleep. Secondly, routing schemes decide on a path in the network between two nodes which are received as input from the application layer via APIs exposed by the routing layer.

Based on ongoing research in sensor networks at Cornell, we propose a two fold extension of the functionality of routing protocols and APIs exposed by them to the applications running on top. First, we propose an interaction between the routing layer and the application layer in which the application is able to decide on a specific subset of nodes in the network that should be involved in routing. The routing layer should then be able to construct paths between the specified set of nodes (multiple sources) and the controller by minimizing the additional nodes used. This should be achieved while preserving the energy efficiency of the routing protocol. Additionally, based on the specified subset of nodes, the routing layer should be capable of setting up these paths dynamically.

Consider the example illustrated in Figure 3. A sensor network is setup such that the controller is located at the top right hand corner. The controller is interested in detecting events in the shaded region (query area) as shown in 2(a). Using techniques proposed in [6], the nodes that lie in the shaded region are

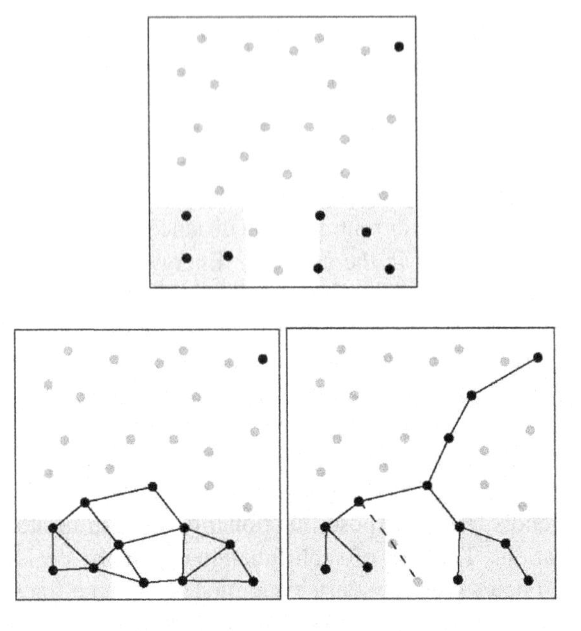

Figure 1.2. Routing tree formation.

determined. The sub-graph in Fig 2(b) represents the subset of nodes that lie in the shaded region and therefore are required to remain awake in order to be able to sense an event and to take part in routing. When forming a routing path from these nodes to the controller node, the number of additional nodes (that lie in the unshaded area) used are minimized as shown in 2(c). The remaining nodes in the network can now be shifted to a low duty cycle mode. In this example, the application running at the controller is required to interact with the routing layer and decide on the subset of nodes that are required to take part in routing. The routing layer should be flexible enough to dynamically form the routing paths based on these constraints. From preliminary results on our research, we have found that this approach leads to a significant amount of energy savings when compared to other energy efficient routing protocols described in [7, 8]. This particular extension has been included in the Tycho system currently being developed at Cornell. Additionally, it is desirable for sensors in the network to uniformly dissipate energy since this guarantees that no single sensor will drain its power prematurely thus possibly disconnecting fractions of the network. The routing layer should bear this responsibility, and it can achieve this by varying the subset of nodes that are sent to sleep during each subsequent sleep cycle.

Various routing protocols to reduce latency in event detection and data aggregation have been proposed. However, it is our belief that if the application is able to negotiate the maximum permissible latency with the routing layer, this will permit a greater degree of flexibility to the routing layer when making decision regarding which nodes should be put into a low duty cycle mode. This is primarily because not all sensor network applications require a low latency network. If an application is able to tolerate high latency in receiving data from an event, the routing layer is then given the freedom to put a large fraction of nodes in the network to sleep. Therefore, when an event is detected by a node, if nodes along the routing path are asleep, data can be cached and sensors can wait till every other sensor on the routing path is awake and then transmits the information to the controller. As a result, the energy saved is a function of the permissible latency. This is the second extension we propose to the interactions between the application and the routing layers.

The interactions described in this section yield more flexibility to the applications and require the various layers to work together cohesively with the goal of saving energy.

4. Position Information in Sensor Networks

Sensor networks by definition sense their surroundings and cannot afford to operate "in the dark" with respect to their position in the field. Position information is necessary for tagging sensor readings [11, 12], geographic routing [10] and caching schemes, clustering and group formation schemes useful in Astrolabe [1], and even addressing the sensors in certain applications [13]. To enable such a wide range of sensor-network applications the community is in search for the perfect sensor-network networking stack and we believe that position services will figure prominently in such a stack. Research in this area has traditionally focused on a very constrictive API for location services, one which consists of a single function that returns a best-estimate point-location for a sensor. This approach is adequate for small-scale static networks where sensor locations can be statically programmed in at deployment time but does not scale well to large or dynamic networks. At Cornell, ongoing research aims to provide a flexible and scalable location-discovery component for the sensor-network networking stack.

As mentioned in the Sextant paper [6] an ideal location discovery protocol would have the following properties:

- **Cheap**: Location discovery should be cheap and consume little power, with minimal dependence on infrastructure in the environment and dedicated hardware on each node.

- **Accurate**: Location discovery should achieve high accuracy. The degree of accuracy should be tunable by the network administrator.

- **Scalable**: The protocol should scale well with increasing number of nodes. Communication load on a node should be independent of the total area of network coverage and the total number of nodes and computation should be distributed evenly.

- **Heterogeneous**: The protocol should support heterogeneous networks where nodes have differing capabilities, such as varying transmission power levels, antenna arrays for determining angle of arrival, configurable angle of transmission and signal strength measurement hardware for relative position estimation.

- **Easy to deploy**: Finally, the protocol should be practical and easy to deploy. Assumptions made in calculating locations should hold in the field.

Sextant, designed with the above goals in mind, necessitates a new API for location services that allows it to return high-fidelity location data and facilitates two-way interactions between Sextant and the applications using it.

A critical issue in location discovery is the representation of a node's position. One approach is to keep and update only a single point estimate for a node. While this approach requires little state, it introduces errors that may compound at the location discovery level as well as at the application level. Sextant, instead, explicitly tracks and refines over time the area a sensor can be located within. To maximize accuracy and minimize storage and communication requirements, it uses Bezier curves to efficiently represent the areas, which need not be convex or even simply connected. Figure 4 depicts nodes along with their Sextant areas. While in its current state Sextant gives applications the guarantee that the sensor resides within the area determined, a simple extension can annotate the returned area with a probability distribution which represents the relative confidence of the system in the node's precise position.

A second issue is the collection of location information. The economic and energy cost of using dedicated positioning hardware, like GPS receivers, at each node is prohibitively high. Instead Sextant infers location information by creating a system of equations, the solution for which gives the area within which each node may be located, and then solves this system in a distributed fashion. Sextant uses the communication hardware already present at each node as a primary source of geographic constraints that it translates into the system of equations. In addition, it can generate additional constraints from other sources including event-sensors and antenna arrays when available. Sextant uses a very small number of landmark nodes that are aware of their own locations, either by static encoding or the use of GPS, to arrive at its solution.

As a power-aware component, Sextant adheres to the guidelines discussed in the previous section by reducing its dependence on power-consuming dedi-

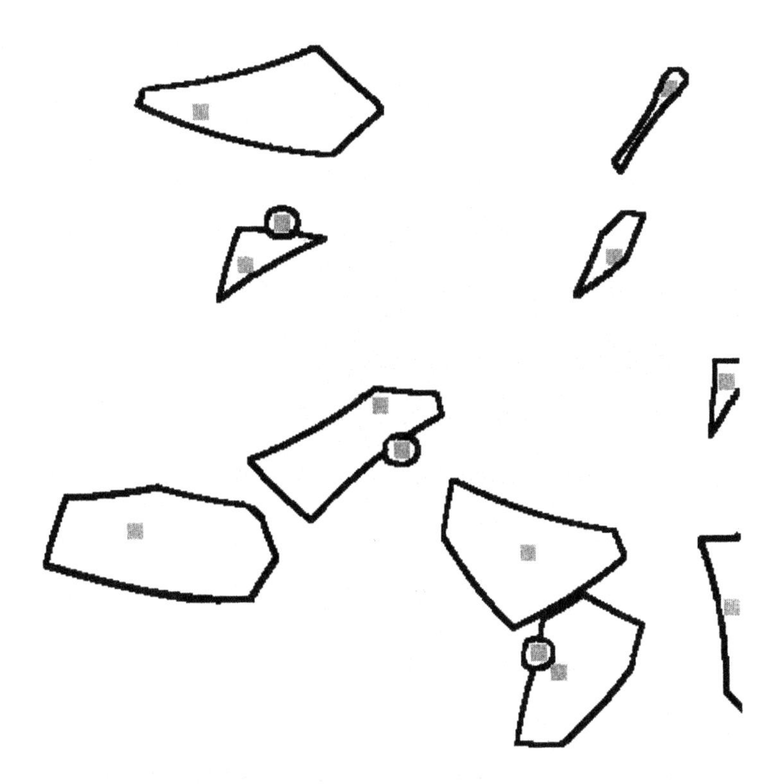

Figure 1.3. Areas determined by Sextant.

cated positioning hardware like GPS receivers and depending instead on MAC level information gleaned from the already present communication hardware. In addition, Sextant converges quickly in static networks obviating the need for constant Sextant-traffic and in highly dynamic networks it limits itself to localized proactive traffic and evenly distributes the processing and communication load leading to uniform energy dissipation in the network.

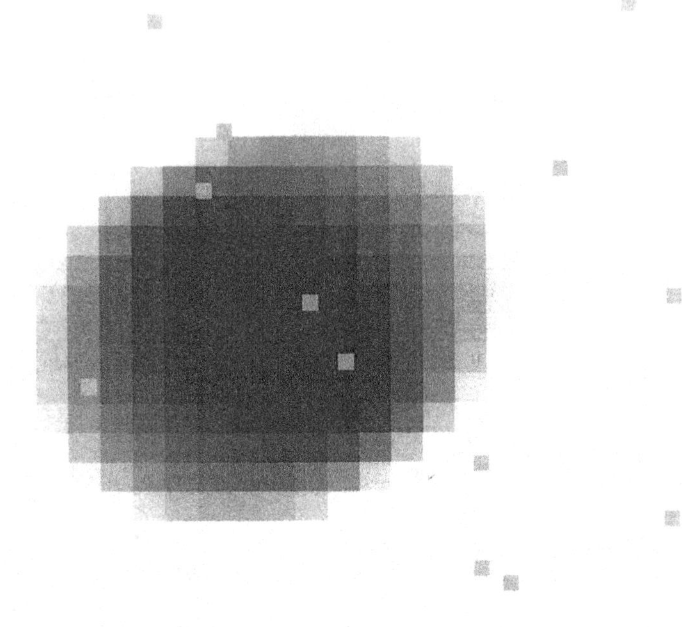

Figure 1.4. Tycho using Sextant areas to assign event detection probabilities.

Of the applications that use the Sextant API, legacy applications can query Sextant for a point-estimate that they are well suited to deal with, while applications that can use the extra information available can do so to minimize their own error. This later approach is evaluated in the Tycho paper [9] where the system, built on top of Sextant, weighs data from different sensors based on the confidence of the sensor's position to produce a probability distribution of an event's location. Figure 1.3 illustrates a node's confidence in the event's location given the Sextant area it lies within. The likelihood of the event taking place in an area is represented using different shades with lighter shades representing low probabilities and darker shades representing high probabilities. Tycho has been shown to be more accurate than traditional triangulation schemes used previously that locate an event to a point-location. In addition, the new API allows Tycho to provide Sextant with additional geographic constraints that it gleans from the already present sensor hardware. This serves to enhance Sextant's location estimates and in turn iteratively increases Tycho's accuracy.

5. Conclusions

Our paper reviewed three technologies matched to the unique needs of pervasive computing environments. Although the components have yet to be integrated into a single platform, doing so is an eventual objective of our effort.

In fact, we believe that the ideas underlying the solutions we present here would also be useful in wired systems. For decades, developers have constructed wired network applications under the assumption that the less each application component "knows" about the network, or about the states of peer components, the better. This sort of thinking is reflected in the prevailing application development models and platforms: client-server systems in their varied forms, Web applications, and most recently the Web Services architecture. One can trace the underlying mindset to the end-to-end philosophy, which can be interpreted as arguing for black-box networks and application designs in which each component is on its own.

It may be time to explore a countervailing view, better matched to the properties of pervasive computing and embedded sensor applications. This view recognizes that the topology of a network, the properties of the components, their positions in the real world and relative to one-another and the constraints under which they operate may have implications for the behavior of other components. Such thinking argues for system services that make it easy for components to share their states and to exploit the information they obtain from one-another to achieve global objectives that would otherwise be unrealizable.

The end-to-end philosophy served us well in developing wired applications, but a new paradigm of sensitivity to system and network state may be needed in response to the unique needs of these new kinds of systems.

References

[1] Astrolabe: A Robust and Scalable Technology for Distributed System Monitoring, Management, and Data Mining. Robbert van Renesse, Kenneth Birman and Werner Vogels. *ACM Transactions on Computer Systems,* May 2003, Vol.21, No. 2, pp 164-206.

[2] Scalable Data Fusion Using Astrolabe. Ken Birman, Robbert van Renesse and Werner Vogels. In proceedings of the Fifth International Conference

on Information Fusion 2002 (IF 2002), July 2002.

[3] Bimodal Multicast. Kenneth P. Birman, Mark Hayden, Oznur Ozkasap, Zhen Xiao, Mihai Budiu and Yaron Minsky. ACM Transactions on Computer Systems, Vol. 17, No. 2, pp 41-88, May, 1999.

[4] Kelips: Building an Efficient and Stable P2P DHT Through Increased Memory and Background Overhead. Indranil Gupta, Ken Birman, Prakash Linga, Al Demers and Robbert van Renesse. Submitted to: 2nd International Workshop on Peer-to-Peer Systems (IPTPS '03); February 20-21, 2003. Claremont Hotel, Berkeley, CA, USA.

[5] J. Hill, R. Szewczyk, A.Woo, S. Hollar, D. E. Culler, and K. S. J. Pister. "System architecture directions for networked sensors", In Proc. the 9th International Conference on Architectural Support for Programming Languages and Operating Systems, Boston, MA, USA, Nov. 2000.

[6] S. Guha and E. G. Sirer, "Distributed Constraint-based Location Discovery in Ad hoc Networks," Cornell University, Tech. Rep. cul.cis/TR2004-1939, 2004.

[7] W. R. Heinzelman, A. Chandrakasan, and H. Balakrishnan, "Energy-Efficient Communication Protocol for Wireless Microsensor Networks," in Proceedings of HICSS, Jan. 2000.

[8] H. Z. Tan and I. Körpeoglu, "Power efficient data gathering and aggregation in wireless sensor networks," ACM SIGMOD Record, vol. 32, no. 4, pp. 66–71, Dec. 2003.

[9] R. Narayan, S. Guha and E. G. Sirer, "Position Informed Energy Efficient Sensing,". Under submission to SECON. 2004.

[10] Brad Karp and H. T. Kung, "GPSR: Greedy Perimeter Stateless Routing for Wireless Networks," in proceedings of International Conference on Mobile Computing and Networking, Aug. 2000.

[11] Janos Sallai, Gyorgy Balogh, Miklos Maroti and Akos Ledeczi, "Acoustic Ranging in Resource Constrained Sensor Networks," Vanderbilt University, Tech. Rep. ISIS-04-504, 2004.

[12] Aram Galstyan, Bhaskar Krishnamachari, Kristina Lerman and Sundeep Pattem, "Distributed Online Localization in Sensor Networks Using a Moving Target," in proceedings of International Symposium on Information Processing in Sensor Networks, Apr. 2004.

[13] Jeremy Elson and Deborah Estrin, "An Address-Free Architecture for Dynamic Sensor Networks," Computer Science Department USC, Tech. Rep. 00-724, 2000.

Chapter 2

ON THE FUNDAMENTAL RELATIONSHIP BETWEEN THE ACHIEVABLE CAPACITY AND DELAY IN MOBILE WIRELESS NETWORKS*

Xiaojun Lin and Ness B. Shroff
School of Electrical and Computer Engineering
Purdue University, West Lafayette, IN 47907
{linx,shroff}@ecn.purdue.edu

Abstract In this work, we establish the fundamental relationship between the achievable capacity and delay in mobile wireless networks. Under an *i.i.d.* mobility model, we first obtain the following upper bound on the achievable capacity given a delay constraint. For a mobile wireless network with n nodes, if the per-bit-averaged mean delay is bounded by $\bar{D}$, then the upper bound on the per-node capacity is on the order of $\sqrt[3]{\frac{\bar{D}}{n}}\log n$. By studying the conditions under which the upper bound is tight, we are able to identify the optimal values of several key scheduling parameters. We then develop a scheduling scheme that can almost achieve the upper bound (up to a logarithmic factor). This suggests that the upper bound is tight. Our scheduling scheme also achieves a provably larger per-node capacity than schemes reported in previous works. In particular, when the delay is bounded by a constant, our scheduling scheme achieves a per-node capacity that is inversely proportional to the cube root of n (up to a logarithmic factor). This implies that, for the *i.i.d.* mobility model, mobility improves the achievable capacity of static wireless networks, even with constant delays! Finally, the insight drawn from the upper bound allows us to identify limiting factors in existing scheduling schemes. These results present a relatively complete picture of the achievable capacity-delay tradeoffs under different settings.

Keywords: Mobile wireless networks, mobile ad hoc networks, capacity-delay tradeoff, large system asymptotics.

*This work has been partially supported by the NSF grants ANI-0207728, EIA-0130599 and the Indiana 21st Century Fund for Wireless Networks.

1. Introduction

Since the seminal paper by Gupta and Kumar [1], there has been tremendous interest in the networking research community to understand the fundamental achievable capacity in wireless networks. For a static network (where nodes do not move), Gupta and Kumar show that the per-node capacity decreases as $O(1/\sqrt{n \log n})$[1] as the number of nodes n increases [1]. The capacity of wireless networks can be improved when *mobility* is taken into account. When the nodes are mobile, Grossglauser and Tse show that per-node capacity of $\Theta(1)$ is achievable [2], which is much better than that of static networks. This capacity improvement is achieved at the cost of excessive packet delays. In fact, it has been pointed out in [2] that the packet delay of the proposed scheme could be unbounded.

There have been several recent studies that attempt to address the relationship between the achievable capacity and the packet delay in mobile wireless networks. In the work by Neely and Modiano [3], it was shown that the maximum achievable per-node capacity of a mobile wireless network is bounded by $O(1)$. Under an *i.i.d.* mobility model, the authors of [3] present a scheme that can achieve $\Theta(1)$ per-node capacity and incur $\Theta(n)$ delay, provided that the load is strictly less than the capacity. Further, they show that it is possible to reduce packet delay if one is willing to sacrifice capacity. In [3], the authors formulate and prove a fundamental tradeoff between the capacity and delay. Let the average end-to-end delay be bounded by D. For D between $\Theta(1)$ and $\Theta(n)$, [3] shows that the maximum per-node capacity λ is upper bounded by

$$\lambda \leq O(\frac{D}{n}). \tag{2.1}$$

The authors of [3] develop schemes that can achieve $\Theta(1)$, $\Theta(1/\sqrt{n})$, and $\Theta(1/(n \log n))$ per-node capacity, when the delay constraint is on the order of $\Theta(n)$, $\Theta(\sqrt{n})$, and $\Theta(\log n)$, respectively.

Inequality (2.1) leads to the *pessimistic* conclusion that a mobile wireless network can sustain at most $O(1/n)$ per-node capacity with a constant delay bound. This capacity is even worse than that of static networks. It turns out that this pessimistic conclusion is due to certain restrictive assumptions that are implicit in the work in [3] (we will elaborate on these assumptions in Section 6). In fact, Toumpis and Goldsmith [4] present a scheme that can achieve a per-node capacity of $\Theta(n^{(d-1)/2}/\log^{5/2} n)$ when the delay is bounded by $O(n^d)$. The result of [4] has incorporated the effect of fading. If we remove fading, the per-node capacity will be of the order $\Theta(n^{(d-1)/2}/\log^{3/2} n)$. Ignoring the logarithmic term, we find that in [4] the following capacity-delay tradeoff is achievable:

$$\lambda^2 = \Theta(\frac{D}{n}). \tag{2.2}$$

This is better than (2.1). In particular, the authors of [4] present a scheme that can achieve $\Theta(1/(\sqrt{n}\log^{3/2} n))$ per-node capacity with a constant delay bound. (The capacity will be $\Theta(1/(\sqrt{n \log n}))$ with no fading.) This capacity is now *comparable* to that of the static wireless networks.

An open question that still remains is: *what is the optimal capacity-delay tradeoff in mobile wireless networks*? Inequality (2.1) is clearly not optimal. The methodology of [4] is constructive in nature. Hence, inequality (2.2) is only a lower bound. The search for the optimal capacity-delay tradeoff is important for two reasons. First, it will allow us to see where the fundamental limits (i.e., upper bounds) are, and how far existing schemes could possibly be improved. Secondly, as has happened in previous works [1, 3], a careful study of the upper bound is usually able to reveal the delicate tradeoffs inherent to the problem. A complete understanding of these tradeoffs will help us identify the possible points of inefficiency in existing schemes and provide directions for further improvement. The ultimate goal is to find a scheme that can achieve the optimal capacity-delay tradeoff.

This paper accomplishes these two goals. Under the *i.i.d.* mobility model studied in [3], we will first establish an *upper bound* on the optimal capacity-delay tradeoff in mobile wireless networks. We will show that, if the per-bit-averaged mean delay is bounded by $\bar{D}$, then the per-node capacity λ is upper bounded by

$$\lambda^3 \leq O(\frac{\bar{D}}{n} \log^3 n). \tag{2.3}$$

In Fig. 2.1, we draw this upper bound alongside the capacity-delay tradeoffs achieved by the schemes in [3] and [4]. The top line corresponds to our upper bound (achievable by the scheme outlined in Section 5 up to a logarithmic factor), the middle (dashed) line is achieved by the scheme in [4], and the bottom (dash-dotted) line is achieved by the scheme in [3]. There is obviously a gap between the upper bound and what can be achieved by existing schemes.

Further, in the process of proving the upper bound, we are able to identify the optimal choices for several key parameters of the scheduling policy. We then develop a new scheme that achieves the upper bound on the capacity-delay tradeoff up to a logarithmic factor, which suggests that our upper bound is fairly tight. Our new scheme achieves a larger per-node capacity than the ones in [3] and [4]. In particular, our scheme can achieve $\Theta(n^{-1/3}/\log n)$ per-node capacity with constant delay. Unlike previous works, this result shows that, even for a constant delay bound, the per-node capacity of mobile wireless networks can be larger than that of the static networks! Finally, the insight drawn from the upper bound allows us to identify the limiting factors of the schemes in [3] and [4].

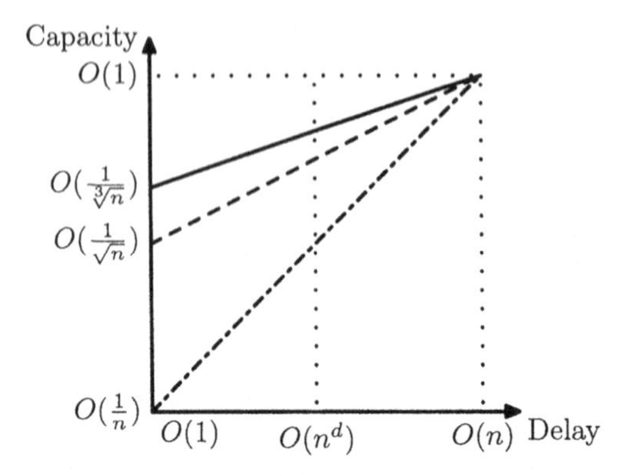

Figure 2.1. The achievable capacity-delay tradeoffs of existing schemes compared with the upper bound (ignoring the logarithmic terms).

The rest of the paper is organized as follows. In Section 2, we outline the network and mobility model. In Section 3, we prove several key properties that capture various tradeoffs inherent in mobile wireless networks. We establish the upper bound on the optimal capacity-delay tradeoff in Section 4 and present a scheme in Section 5 that achieves a capacity-delay tradeoff close to the upper bound. In Section 6, we discuss the existing schemes described in [3] and [4]. Then we conclude.

2. Network and Mobility Model

We consider a mobile wireless network with n nodes moving within a unit square[2]. We assume that time is divided into slots of unit length. We assume the following *i.i.d. mobility model* proposed in [3]. At each time slot, the positions of each node are *i.i.d.* and uniformly distributed within the unit square. Between time slots, the distributions of the positions of the nodes are independent. Although the assumption on an *i.i.d.* mobility model is somewhat restrictive, its mathematical tractability allows us to gain important insights into the structure of the problem. We will comment on some extensions to the *i.i.d.* mobility model in the conclusion.

For simplicity, we assume the following traffic model similar to the models in [3, 4]. We assume that the number of nodes n is even and the nodes can be labeled in such a way that node $2i-1$ communicates with node $2i$, and node $2i$ communicates with node $2i-1, i = 1, 2, ..., n/2$. The communication between any source-destination pairs can go through multiple other nodes as *relays*. That is, the *source* can either send a message directly to the *destination*; or, it can send the message to one or more *relay* nodes; the relay nodes can further forward the message to other relay nodes (possibly after moving to another position); and finally some relay node forwards the message to the destination.

We assume the following Protocol Model from [1] that governs *direct radio transmissions* between nodes. Let W be the bandwidth of the system. Let X_i denote the position of node i, $i = 1, ..., n$. Let $|X_i - X_j|$ be the Euclidean distance between nodes i and j. At each time slot, node i can communicate *directly* with another node j at W bits per second if and only if the following interference constraint is satisfied [1]:

$$|X_j - X_k| \geq (1 + \Delta)|X_i - X_j|$$

for every other node $k \neq i, j$ that is simultaneously transmitting. Here, Δ is some positive number. Note that an alternative model for direct radio transmission is the Physical Model [1, 4]. In the Physical Model, a node can communicate with another node if the signal-to-interference ratio is above a given threshold. It has been shown that, under certain conditions, the Physical Model can be reduced to the Protocol Model with an appropriate choice of Δ [1]. Hence, we will not consider the Physical Model any further in this paper. We also assume that no nodes can transmit and receive over the same frequency at the same time. We further assume the following *separation of time scale*, i.e., radio transmission can be scheduled at a time scale much faster than that of node mobility. This is usually a reasonable assumption in real networks. Hence, a message may be divided into multiple bits and each bit can be forwarded multiple hops separately within a single time slot.

We assume a uniform traffic pattern, that is, all source nodes communicate with their destination nodes at the same rate λ. let $\bar{D}$ be the mean delay averaged over all messages and all source-destination pairs. Both λ and $\bar{D}$ will depend on how the transmissions between mobile nodes are scheduled. We are interested in capturing the fundamental tradeoff between the achievable capacity λ and the delay $\bar{D}$. That is, over all possible ways of scheduling the radio transmissions, what is the maximum per-node capacity λ given certain constraint on the delay $\bar{D}$.

3. Properties of the Scheduling Policies

In this section, we will prove several key results that capture the various tradeoffs inherent in mobile wireless networks. We will first define the class of scheduling policies that we will consider. Because we are interested in the fundamental *achievable* capacity for a given delay, we will assume that there exists a scheduler that has all the information about the current and past status of the network, and can schedule any radio transmission in the current and future time slots. At each time slot t, for each bit b that has not been delivered to its destination yet, the scheduler needs to perform the following two functions:

- *Capture:* The scheduler needs to decide whether to deliver the bit b to the destination within the current time slot. If yes, the scheduler then needs to choose one relay node (possibly the source) that has a copy of the bit b at the beginning of the time slot t, and schedule radio transmissions to forward this bit to the destination *within the same time slot*, using possibly multi-hop transmissions. When this happens successfully, we say that the chosen relay node has successfully *captured* the destination of bit b. It is important to forward the bit to the destination within a single time slot. Otherwise, since the chosen relay node may move far away from the destination in the next time slot, the nodes that received the bit b in the current time slot will only count as new relay nodes for the bit b, and they have to capture again in the next time slot.

- *Duplication:* If capture does not occur for bit b, the scheduler needs to decide whether to *duplicate* bit b to other nodes that do not have the bit at the beginning of the time slot t. The scheduler also needs to decide which nodes to relay from and relay to, and how to schedule radio transmissions to forward the bit to these new relay nodes.

In this paper, we will consider the class of *causal* scheduling policies that perform the above two functions at each time slot. The causality assumption essentially requires that, when the scheduler makes the capture decision and the duplication decision, it can only use information about the current and the past status of the network. In particular, at any time slot t, the scheduler cannot use information about the *future* positions of the nodes at any time slot $s > t$.

This class of scheduling policies is clearly very general, and encompasses nearly any practical scheduling scheme we can think of. (Note that even *predictive* scheduling schemes have to rely on current and past information only.) Some remarks on the *capture* process is in order. Although we do allow for other less intuitive alternatives, in a typical scheduling policy a successful *capture* usually occurs when some relay nodes are within an area close to the destination node, so that fewer resources will be needed to forward the information to the destination. For example, a relay node could enter a disk of a certain radius around the destination, or a relay node could enter the same cell as the destination. We call such an area a *capture neighborhood.* The relay nodes that has the bit b at the beginning of the time slot t are called *mobile relays* for bit b. The mobile relay that is chosen to forward the bit b to the destination is called the *last mobile relay* for bit b.

The following examples are illustrative of the possible scheduling policies within this broad class. The schemes in previous works [3, 4] are all special cases or variants of these examples.

Example A: The number of mobile relays R is fixed and the capture neighborhood is chosen to be a disk with a fixed radius ρ around the destination.

Once a bit b enters the system, it is immediately broadcast to the nearest $R-1$ neighboring nodes. When any of the R mobile relays (including the source node) move within distance ρ from the destination, the bit b is then forwarded from the nearest mobile relay to the destination.

Example B: The unit area is divided into a number of cells. Once a bit b enters the system, it is immediately broadcast to all other nodes in the same cell. The number of mobile relays for the bit b then stay unchanged. Note that the actual number of mobile relays depends on the number of nodes that reside in the same cell as the source (at the time slot when the bit b enters the system), and is thus a random variable. When one of the mobile relays moves into the same cell as the destination, the bit b is then forwarded from the nearest mobile relay to the destination.

Example C: In the above two schemes, no duplication for bit b is carried out except at the first time slot when the bit enters the system. A more sophisticated strategy is to use an "opportunistic duplication scheme" such as the example below. The unit area is divided into a number of cells. After a bit b enters the system, at each time slot t, if one of the mobile relays moves into the same cell as the destination, bit b is then forwarded from the nearest mobile relay to the destination. Otherwise, the source node (or, alternatively, the current mobile relays) broadcasts the bit to all other nodes that reside at the same cell. Hence, duplication may occur at each time slot until bit b is delivered to its destination.

In the sequel, we will prove several key inequalities that capture the various tradeoffs inherent in this broad class of scheduling policies. Intuitively, the larger the number of mobile relays and the larger the capture neighborhood, the smaller the delay. On the other hand, in order to improve capacity, we need to consume fewer radio resources, which implies a smaller number of mobile relays and a shorter distance from the last mobile relay to the destination. As we will see later, these tradeoffs will determine the fundamental relationship between achievable capacity and delay in mobile wireless networks.

3.1 Notations

Let $(\Omega, \mathcal{F}, P)$ be the probability space on which the random mobility of the mobile nodes is defined. Let $X(i,t)$ be the random variable that denotes the position of node i at time slot t. Let b denote a bit that needs to be communicated from a source node $S(b)$ to destination node $D(b)$. Let $t_0(b)$ be the time slot when bit b first enters the system. Let $I_b(i,t)$ be an indicator function, where $I_b(i,t)=1$ if node i has a copy of bit b at the beginning of time slot t, $I_b(i,t)=0$ otherwise. By definition, $I_b(S(b), t_0(b)) = 1$, and $I_b(i,t) = 0$ for all i and $t < t_0(b)$. Let $\mathcal{F}_t$ be the σ-algebra generated by the random variables $X(i,s)$ and $I_b(i,s)$ for all $s \le t$. Hence $\{\mathcal{F}_t, t = 0, 1, ...\}$ is a filtration [5, p231] and $\mathcal{F}_t$ captures all information about the "history" up to time slot t.

Fix any scheduling policy and fix a bit b that enters the system at time slot $t_0(b)$. For any time slot $t \geq t_0(b)$, let $C_b(t) = 1$ if the scheduler decides that a successful capture occurs at this time slot. $C_b(t) = 0$, otherwise. If $C_b(t) = 1$, the scheduler then picks one mobile relay that has a copy of the bit b at the beginning of the time slot to forward the bit towards the destination *within the same time slot* t, using possibly multi-hop transmissions. Let $\tilde{l}_b(t)$ be the distance from the chosen mobile relay to the destination of the bit b. Let $\tilde{l}_b(t) = \infty$ if $C_b(t) = 0$. Finally, let $r_b(t+1)$ denote the number of mobile relays holding the bit b *at the end* of the time slot t, i.e., $r_b(t+1)$ is the cardinality of the set $\{i : I_b(i, t+1) = 1\}$. Since the random variables $C_b(t), \tilde{l}_b(t)$ and $r_b(t+1)$ are all outcomes of the scheduling policy, the causality assumption implies that they are all $\mathcal{F}_t$-measurable[3].

Let

$$s_b \triangleq \min\{t : t \geq t_0(b) \text{ and } C_b(t) = 1\}$$

be the first time when a successful capture for bit b occurs. Thus s_b is a stopping time [5, p234] with respect to the filtration $\{\mathcal{F}_t, t = 0, 1, ...\}$. Let $R_b \triangleq r_b(s_b)$ denote the number of mobile relays holding the bit b at the time of capture. Let $D_b \triangleq s_b - t_0(b)$ denote the number of time slots from the time bit b enters the system to the time of capture. Let $l_b \triangleq \tilde{l}_b(s_b)$ denote the distance from the chosen last mobile relay node to the destination. The quantities R_b, D_b, and l_b are essential for the tradeoffs that follow. Note that D_b includes possible queuing delay at the source node or at the relay nodes.

3.2 Tradeoff I : D_b versus R_b and l_b

PROPOSITION 2.1 *Under the* i.i.d. *mobility model, the following inequality holds for any causal scheduling policy when* $n \geq 3$,

$$c_1 \log n \mathbf{E}[D_b] \geq \frac{1}{(\mathbf{E}[l_b] + \frac{1}{n^2})^2 \mathbf{E}[R_b]} \text{ for all bits } b, \tag{2.4}$$

where c_1 *is a positive constant.*

The proof is available in Appendix 2.A. This new result is one of the cornerstones for deriving the optimal capacity-delay tradeoff in mobile wireless networks. It captures the following tradeoff: the smaller the number R_b of mobile relays the bit b is duplicated to, and the shorter the targeted distance l_b from the last mobile relay to the destination, the longer it takes to capture the destination. This seemingly odd relationship is actually motivated by some simple examples. Consider Example A at the beginning of Section 3. When R_b and the area of the capture neighborhood A_b are constants, then $1 - (1 - A_b)^{R_b}$ is the probability that any one out of the R_b nodes can capture the destination in one time slot. It is easy to show that, the average number of time slots needed before a successful capture occurs, is,

$$\mathbf{E}[D_b] = \frac{1}{1-(1-A_b)^{R_b}} \geq \frac{1}{A_b R_b}.$$

If, as in Example B, R_b and possibly A_b are random but *fixed after the first time slot* $t_0(b)$, then

$$\mathbf{E}[D_b | R_b, A_b] \geq \frac{1}{A_b R_b}.$$

By Hőlder's Inequality [5, p15],

$$\mathbf{E}^2[\frac{1}{\sqrt{A_b}}] \leq \mathbf{E}[R_b]\mathbf{E}[\frac{1}{A_b R_b}].$$

Hence,

$$\begin{aligned} \mathbf{E}[D_b] &\geq \mathbf{E}[\frac{1}{A_b R_b}] \geq \mathbf{E}^2[\frac{1}{\sqrt{A_b}}]\frac{1}{\mathbf{E}[R_b]} \\ &\geq \frac{1}{\mathbf{E}^2[\sqrt{A_b}]\mathbf{E}[R_b]}, \end{aligned}$$

where in the last step we have applied Jensen's Inequality [5, p14]. Note that on average l_b is on the order of $\sqrt{A_b}$. Hence,

$$\mathbf{E}[D_b] \geq \frac{c_1'}{\mathbf{E}^2[l_b]\mathbf{E}[R_b]} \text{ for all bits } b, \tag{2.5}$$

where c_1' is a positive constant. It may appear that, when an "opportunistic duplication scheme" such as the one in Example C is employed, such a scheme might achieve a better tradeoff than (2.5) by starting off with fewer mobile relays and a smaller capture neighborhood, if the node positions at the early time slots after the bit's arrival turns out to be favorable. However, Proposition 2.1 shows that no scheduling policy can improve the tradeoff by more than a $\log n$ factor. For details, please refer to Appendix 2.A.

3.3 Tradeoff II : Multihop

Once a successful capture occurs, the chosen mobile relay (i.e., the *last mobile relay*) will start transmitting the bit to the destination *within a single time slot*, using possibly other nodes as relays. We will refer to these latter relay nodes as *static relays*. The static relays are only used for forwarding the bit to the destination *after a successful capture occurs*. Let h_b be the number of hops it takes from the last mobile relay to the destination. Let S_b^h denote the transmission range of each hop $h = 1, .., h_b$. The following relationship is trivial.

PROPOSITION 2.2 *The sum of the transmission ranges of the h_b hops must be no smaller than the straight-line distance from the last mobile relay to the destination, i.e.,*

$$\sum_{h=1}^{h_b} S_b^h \geq l_b. \tag{2.6}$$

3.4 Tradeoff III : Radio Resources

It consumes radio resources to duplicate each bit to mobile relays and to forward the bit to the destination. Proposition 2.3 below captures the following tradeoff: the larger the number of mobile relays R_b and the further the multi-hop transmissions towards the destination have to traverse, the smaller the achievable capacity. Consider a large enough time interval T. The total number of bits communicated end-to-end between all source-destination pairs is λnT.

PROPOSITION 2.3 *Assume that there exist positive numbers c_2 and N_0 such that $D_b \leq c_2 n^2$ for $n \geq N_0$. If the positions of the nodes within a time slot are* i.i.d. *and uniformly distributed within the unit square, then there exist positive numbers N_1 and c_3 that only depend on c_2, N_0 and Δ, such that the following inequality holds for any causal scheduling policy when $n \geq N_1$,*

$$\sum_{b=1}^{\lambda nT} \frac{\Delta^2}{4} \frac{\mathbf{E}[R_b] - 1}{n} + \mathbf{E}[\sum_{b=1}^{\lambda nT} \sum_{h=1}^{h_b} \frac{\pi \Delta^2}{4} (S_b^h)^2] \leq c_3 WT \log n. \tag{2.7}$$

The assumption that $D_b \leq c_2 n^2$ for large n is not as restrictive as it appears. It has been shown in [3] that the maximal achievable per-node capacity is $\Theta(1)$ and this capacity can be achieved with $\Theta(n)$ delay. Hence, we are most interested in the case when the delay is not much larger than the order $O(n)$. Further, Proposition 2.3 only requires that the stationary distribution of the positions of the nodes within a time slot is *i.i.d.* It does not require the distribution between time slots to be independent.

We briefly outline the motivation behind the inequality (2.7). The details of the proof are quite technical and available in Appendix 2.B. Consider nodes i, j that directly transmit to nodes k and l, respectively, at the same time. Then, according to the interference constraint:

$$\begin{aligned} |X_j - X_k| &\geq (1+\Delta)|X_i - X_k] \\ |X_i - X_l| &\geq (1+\Delta)|X_j - X_l]. \end{aligned}$$

Hence,

$$\begin{aligned}|X_j - X_i| &\geq |X_j - X_k| - |X_i - X_k| \\ &\geq \Delta|X_i - X_k|.\end{aligned}$$

Similarly,

$$|X_i - X_j| \geq \Delta|X_j - X_l|.$$

Therefore,

$$|X_i - X_j| \geq \frac{\Delta}{2}(|X_i - X_k| + |X_j - X_l|).$$

That is, disks of radius $\frac{\Delta}{2}$ times the transmission range centered at the transmitter are disjoint from each other[4]. This property can be generalized to *broadcast* as well. We only need to define the transmission range of a broadcast as the distance from the transmitter to the furthest node that can successfully receive the bit. The above property motivates us to measure the radio resources each transmission consumes by the areas of these disjoint disks [1]. For unicast transmissions from the last mobile relay to the destination, the area consumed by each hop is $\frac{\pi\Delta^2}{4}(S_b^h)^2$. For duplication to other nodes, broadcast is more beneficial since it consumes fewer resources. Assume that each transmitter chooses the transmission range of the broadcast independently of the positions of its neighboring nodes. If the transmission range is s, then on average no greater than $n\pi s^2$ nodes can receive the broadcast, and a disk of radius $\frac{\Delta}{2}s$ (i.e., area $\frac{\pi\Delta^2}{4}s^2$) centered at the transmitter will be disjoint from other disks. Therefore, we can use $\frac{\Delta^2}{4}\frac{\mathbf{E}[R_b]-1}{n}$ as a lower bound on the expected area consumed by duplicating the bit to $R_b - 1$ mobile relays (excluding the source node). This lower bound will hold even if the duplication process is carried out over multiple time slots, because the average number of *new* mobile relays each broadcast can cover is at most proportional to the area consumed by the broadcast. Therefore, inspired by [1], the amount of radio resources consumed must satisfy

$$\sum_{b=1}^{\lambda nT} \frac{\Delta^2}{4}\frac{\mathbf{E}[R_b]-1}{n} + \mathbf{E}[\sum_{b=1}^{\lambda nT}\sum_{h=1}^{h_b}\frac{\pi\Delta^2}{4}(S_b^h)^2] \leq c_3' WT, \tag{2.8}$$

where c_3' is a positive constant.

However, $\frac{\Delta^2}{4}\frac{\mathbf{E}[R_b]-1}{n}$ may fail to be a lower bound on the expected area consumed by duplicating to $R_b - 1$ mobile relays if the following *opportunistic broadcast scheme* is used. The source may choose to broadcast *only when there are a larger number of nodes close by.* If the source can afford to wait for these "good opportunities", an *opportunistic broadcast scheme* may consume less radio resources than a non-opportunistic scheme to duplicate the bit to the same number of mobile relays. Nonetheless, Proposition 2.3 shows that no scheduling policies can improve the tradeoff by more than a $\log n$ factor. For details, please refer to Appendix 2.B.

3.5 Tradeoff IV : Half Duplex

Finally, since we assume that no node can transmit and receive over the same frequency at the same time (a practically necessary assumption for most wireless devices), the following property can be shown as in [1].

PROPOSITION 2.4 *The following inequality holds,*

$$\sum_{b=1}^{\lambda nT} \sum_{h=1}^{h_b} 1 \leq \frac{WT}{2} n. \tag{2.9}$$

4. The Upper Bound on the Capacity-Delay Tradeoff

Our first main result is to derive, from the above four tradeoffs, the upper bound on the optimal capacity-delay tradeoff of mobile wireless networks under the *i.i.d.* mobility model. Since the maximal achievable per-node capacity is $\Theta(1)$ and this capacity can be achieved with $\Theta(n)$ delay by the scheme of [3], we are only interested in the case when the mean delay is $o(n)$.

PROPOSITION 2.5 *Let $\bar{D}$ be the mean delay averaged over all bits and all source-destination pairs, and let λ be the throughput of each source-destination pair. If $\bar{D} = O(n^d), 0 \leq d < 1$, the following upper bound holds for any causal scheduling policy,*

$$\lambda^3 \leq O(\frac{\bar{D}}{n} \log^3 n).$$

Proof: Using the Cauchy-Schwartz inequality, we have

$$\begin{aligned} \left(\sum_{b=1}^{\lambda nT} \sum_{h=1}^{h_b} S_b^h \right)^2 &\leq \left(\sum_{b=1}^{\lambda nT} \sum_{h=1}^{h_b} 1 \right) \left(\sum_{b=1}^{\lambda nT} \sum_{h=1}^{h_b} (S_b^h)^2 \right) \\ &\leq \frac{WTn}{2} \sum_{b=1}^{\lambda nT} \sum_{h=1}^{h_b} (S_b^h)^2, \end{aligned} \tag{2.10}$$

where in the last step we have used Tradeoff IV (2.9). Equality holds in (2.10) when inequality (2.9) is tight and when S_b^h is equal for all b and h. We thus have,

$$\mathbf{E}[\sum_{b=1}^{\lambda nT} \sum_{h=1}^{h_b} (S_b^h)^2] \geq \frac{2}{WTn} \mathbf{E}[\left(\sum_{b=1}^{\lambda nT} \sum_{h=1}^{h_b} S_b^h \right)^2]$$

$$\geq \frac{2}{WTn}\left(\mathbf{E}[\sum_{b=1}^{\lambda nT}\sum_{h=1}^{h_b} S_b^h]\right)^2 \tag{2.11}$$

$$\geq \frac{2}{WTn}\left(\sum_{b=1}^{\lambda nT}\mathbf{E}[l_b]\right)^2, \tag{2.12}$$

where in the last two steps we have used Jensen's Inequality and the Tradeoff II (2.6), respectively. Inequality (2.11) is tight when $\sum_{b=1}^{\lambda nT}\sum_{h=1}^{h_b} S_b^h$ is almost surely a constant, and (2.12) is tight when (2.6) is tight.

From Tradeoff I (2.4), we have

$$\sum_{b=1}^{\lambda nT}\mathbf{E}[R_b] \geq \sum_{b=1}^{\lambda nT}\frac{1}{c_1\log n}\frac{1}{(\mathbf{E}[l_b]+\frac{1}{n^2})^2\mathbf{E}[D_b]}. \tag{2.13}$$

Let

$$\bar{D} = \frac{\sum_{b=1}^{\lambda nT}\mathbf{E}[D_b]}{\sum_{b=1}^{\lambda nT} 1} = \frac{\sum_{b=1}^{\lambda nT}\mathbf{E}[D_b]}{\lambda nT}.$$

Using Jensen's Inequality and Hőlder's Inequality, we have,

$$\frac{1}{\left(\frac{\sum_{b=1}^{\lambda nT}(\mathbf{E}[l_b]+\frac{1}{n^2})}{\sum_{b=1}^{\lambda nT} 1}\right)^2} \leq \left(\frac{\sum_{b=1}^{\lambda nT}\frac{1}{(\mathbf{E}[l_b]+\frac{1}{n^2})}}{\sum_{b=1}^{\lambda nT} 1}\right)^2$$

$$\leq \frac{\sum_{b=1}^{\lambda nT}\frac{1}{(\mathbf{E}[l_b]+\frac{1}{n^2})^2\mathbf{E}[D_b]}}{\sum_{b=1}^{\lambda nT} 1}\frac{\sum_{b=1}^{\lambda nT}\mathbf{E}[D_b]}{\sum_{b=1}^{\lambda nT} 1}. \tag{2.14}$$

Equality holds when $\mathbf{E}[l_b]$ is the same for all b and $\mathbf{E}[D_b] = \bar{D}$ for all b. Substituting (2.14) in (2.13), we have

$$\sum_{b=1}^{\lambda nT}\mathbf{E}[R_b] \geq \frac{1}{c_1\log n}\frac{\left(\sum_{b=1}^{\lambda nT} 1\right)^3}{\bar{D}\left(\sum_{b=1}^{\lambda nT}(\mathbf{E}[l_b]+\frac{1}{n^2})\right)^2}. \tag{2.15}$$

Substituting (2.12) and (2.15) into Tradeoff III (2.7), we have

$$\begin{aligned}\frac{4c_3 WT\log n}{\Delta^2} &\geq \sum_{b=1}^{\lambda nT}\frac{\mathbf{E}[R_b]-1}{n}+\pi\mathbf{E}[\sum_{b=1}^{\lambda nT}\sum_{h=1}^{h_b}(S_b^h)^2] \\ &\geq \frac{1}{c_1 n\log n}\frac{(\lambda nT)^3}{\bar{D}\left(\sum_{b=1}^{\lambda nT}(\mathbf{E}[l_b]+\frac{1}{n^2})\right)^2} \\ &\quad +\frac{2\pi}{WTn}\left(\sum_{b=1}^{\lambda nT}\mathbf{E}[l_b]\right)^2-\lambda T.\end{aligned}$$

There are two cases that we need to consider.

Case 1: If $\sum_{b=1}^{\lambda nT}\mathbf{E}[l_b]\leq\frac{\lambda T}{n}$, then

$$\begin{aligned}\frac{4c_3 WT\log n}{\Delta^2} &\geq \frac{1}{c_1 n\log n}\frac{(\lambda nT)^3}{\bar{D}\left(\frac{2\lambda T}{n}\right)^2}-\lambda T \\ &= \frac{1}{4c_1\log n}\frac{\lambda Tn^4}{\bar{D}}-\lambda T.\end{aligned}$$

When $\bar{D}=O(n^d), d<1$, the first term dominates when n is large. Hence, for n large enough,

$$\begin{aligned}\frac{4c_3 WT\log n}{\Delta^2} &\geq \frac{1}{8c_1\log n}\frac{\lambda Tn^4}{\bar{D}} \\ \lambda &\leq \frac{32c_1c_3W}{\Delta^2}\frac{\bar{D}\log^2 n}{n^4}. \qquad (2.16)\end{aligned}$$

Case 2: If $\sum_{b=1}^{\lambda nT}\mathbf{E}[l_b]\geq\frac{\lambda T}{n}$, then

$$\begin{aligned}\frac{4c_3 WT\log n}{\Delta^2} &\geq \frac{1}{c_1 n\log n}\frac{(\lambda nT)^3}{\bar{D}\left(2\sum_{b=1}^{\lambda nT}\mathbf{E}[l_b]\right)^2} \\ &\quad +\frac{2\pi}{WTn}\left(\sum_{b=1}^{\lambda nT}\mathbf{E}[l_b]\right)^2-\lambda T \qquad (2.17) \\ &\geq 2\sqrt{\frac{1}{c_1\log n}\frac{2\pi}{WTn^2}\frac{(\lambda nT)^3}{4\bar{D}}}-\lambda T \qquad (2.18) \\ &= 2\sqrt{\frac{\pi}{2c_1\log n}\frac{\lambda^3 nT^2}{\bar{D}W}}-\lambda T. \qquad (2.19)\end{aligned}$$

Therefore, either

$$\lambda \leq O(\frac{\bar{D}\log n}{n}), \tag{2.20}$$

or, if $\lambda = \omega(\frac{\bar{D}\log n}{n})$, then the first term in (2.19) dominates when n is large. In the latter case, for n large enough,

$$\begin{aligned} \frac{4c_3 WT\log n}{\Delta^2} &\geq \sqrt{\frac{\pi}{2c_1\log n}\frac{\lambda^3 nT^2}{\bar{D}W}} \\ \lambda^3 &\leq \frac{32c_1c_3^2W^3}{\Delta^4}\frac{\bar{D}\log^3 n}{n}. \end{aligned} \tag{2.21}$$

Finally, we compare the three inequalities we have obtained, i.e., (2.16), (2.20) and (2.21). Since $\bar{D} = o(n^d), d < 1$, inequality (2.21) will eventually be the loosest for large n. Hence, the optimal capacity-delay tradeoff is upper bounded by

$$\lambda^3 \leq O(\frac{\bar{D}}{n}\log^3 n).$$

Q.E.D.

5. An Achievable Lower Bound on the Capacity-Delay Tradeoff

The capacity-delay tradeoff in Proposition 2.5 is better than those reported in [3] and [4]. Assuming that the delay bound is $\Theta(n^d)$, $0 \leq d < 1$, the achievable per-node capacity is $O(n^{-(1-d)})$ by the scheme in [3], and $O(n^{-(1-d)/2})$ by the scheme in [4]. Our upper bound, however, implies a per-node capacity of $O(n^{-(1-d)/3})$ (we have ignored all $\log n$ factors). Since $d < 1$, there is clearly room to substantially improve existing schemes (see Fig. 2.1).

In this section, we will show how the study of the upper bound also helps us to develop a new scheme that can achieve a capacity-delay tradeoff that is close to the upper bound. Precisely, we met several inequalities (2.10)-(2.18) during the derivation of the upper bound. By studying the conditions under which these inequalities are tight, we will be able to identify the optimal choices of various key parameters of the scheduling policy. In the end, the knowledge of the optimal choices of the parameters will help us develop a new scheme that is superior to existing ones.

5.1 Choosing the Optimal Values of the Key Parameters

Assume that the mean delay is bounded by $n^d, d < 1$. By Proposition 2.5, we have,

$$\lambda \leq \Theta(\sqrt[3]{\frac{\bar{D}}{n}\log^3 n}) = \Theta(n^{\frac{d-1}{3}}\log n).$$

In order to achieve the maximum capacity on the right hand side, all inequalities (2.10)-(2.18) should hold with equality. By checking the conditions when (2.10)-(2.14) are tight, we can infer that the parameters (such as $S_b^h, \mathbf{E}[l_b], \mathbf{E}[D_b]$) of each bit b should be about the same and should concentrate on their respective average values. This implies that the scheduling policy should use the same parameters for all bits. From now on, we will assume that all key parameters (such as R_b, l_b, etc.) are indeed the same for all bits.

The inequality (2.18) is essential for deriving the optimal values of these parameters. Note that equality holds in (2.18) if and only if

$$\frac{1}{4c_1 n\log n}\frac{(\lambda nT)^3}{\bar{D}(\sum_{b=1}^{\lambda nT}\mathbf{E}[l_b])^2} = \frac{2\pi}{WTn}(\sum_{b=1}^{\lambda nT}\mathbf{E}[l_b])^2.$$

Substituting $\sum_{b=1}^{\lambda nT}\mathbf{E}[l_b] = \lambda nTl_b$, we can solve for l_b,

$$\begin{aligned}\frac{1}{4c_1 n\log n}\frac{\lambda nT}{\bar{D}l_b^2} &= \frac{2\pi}{WTn}(\lambda nT)^2 l_b^2\\ l_b^4 &= \frac{1}{8\pi c_1}\frac{W}{\bar{D}\lambda n\log n}.\end{aligned}$$

Substituting $\lambda = \Theta(n^{(d-1)/3}\log n)$ and $\bar{D} = n^d$, we obtain the optimal value of l_b,

$$l_b = \Theta(n^{-\frac{1+2d}{6}}\log^{-\frac{1}{2}} n)$$

A reasonable choice for the area of capture neighborhood, A_b, is then,

$$A_b = l_b^2 = \Theta(n^{-\frac{1+2d}{3}}/\log n).$$

By setting (2.9) of Tradeoff IV to equality, we have

$$\begin{aligned}\lambda nTh_b &= \frac{WTn}{2}\\ h_b &= \frac{W}{2\lambda} = \Theta(n^{\frac{1-d}{3}}/\log n).\end{aligned}$$

Table 2.1. The order of the optimal values of the parameters when the mean delay is bounded by n^d.

R_b: # of Duplicates	$\Theta(n^{(1-d)/3})$
l_b: Distance to Destination	$\Theta(n^{-(1+2d)/6}/\log^{1/2} n)$
h_b: # of Hops	$\Theta(n^{(1-d)/3}/\log n)$
S_b^h: Transmission Range of Each Hop	$\Theta(\sqrt{\frac{\log n}{n}})$

By setting (2.6) of Tradeoff II to equality, we have

$$S_b^h = \frac{l_b}{h_b} = \Theta(\sqrt{\frac{\log n}{n}}).$$

Finally, by setting (2.4) of Tradeoff I to equality, we have

$$R_b = \Theta(\frac{1}{c_1 \log n} \frac{1}{l_b^2 \bar{D}}) = \Theta(n^{\frac{1-d}{3}}).$$

The optimal values of these parameters are summarized in Table 2.1.

Several remarks are in order. Since it is sufficient to control all parameters around these optimal values, simple cell-based schemes such as the one in Example B of Section 3 suffice. Secondly, the optimal values for R_b and l_b can provide guidelines on how to choose the cell partitioning. Thirdly, the optimal value for S_b^h is roughly the average distance between neighboring nodes when n nodes are uniformly distributed in a unit square. Hence, it is desirable to use multi-hop transmission over neighboring nodes to forward the information from the last mobile relay to the destination. These guidelines have sketched a blueprint of the optimal scheduling scheme for us. We next present schemes that can achieve capacity-delay tradeoffs that are close to the upper bound up to a logarithmic factor.

5.2 Achievable Capacity with $\Theta(1)$ Delay

We first focus on the case when the mean delay is bounded by a constant, i.e., the exponent $d = 0$. By Proposition 2.5, the per-node throughput is bounded by $O(n^{-1/3} \log n)$. We now present a scheme that can achieve $\Theta(n^{-1/3}/\log n)$ capacity with $\Theta(1)$ delay for large n. This is an encouraging result for mobile networks because we know that the per-node capacity of static networks is $O(1/\sqrt{n \log n})$ [1]. Hence, mobility increases the capacity even with constant delay.

We will need the following Lemma before stating the main scheduling scheme. We will repeatedly use the following type of cell-partitioning. Let m be a positive integer. Divide the unit square into $m \times m$ cells (in m rows and

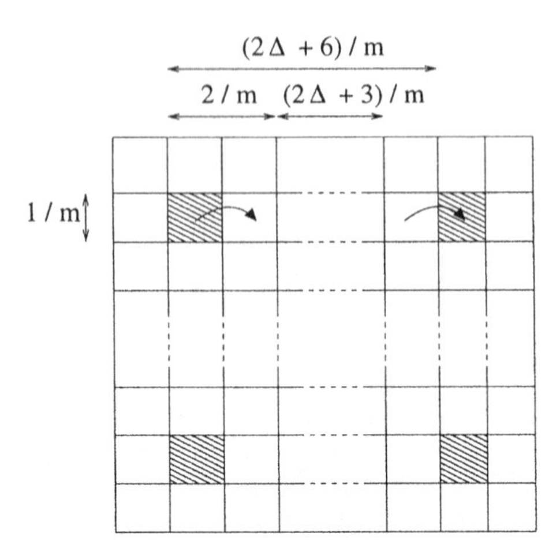

Figure 2.2. Cells that are $\lfloor 2\Delta + 6 \rfloor / m$ apart (i.e., the shaded cells in the figure) can be active together.

m columns, see Fig. 2.2). Each cell is a square of area $1/m^2$. As in [4], we call two cells *neighbors* if they share a common boundary, and we call two nodes *neighbors* if they lie in the same or neighboring cells. We say that a group of cells can be *active* at the same time when one node in each cell can successfully transmit to or receive from a neighboring node, subject to the interference from other cells that are active at the same time. Let $\lfloor x \rfloor$ be the largest integer smaller than or equal to x. The proof of the following Lemma is available in Appendix 2.C.

LEMMA 2.6 *There exists a scheduling policy such that each cell can be active for at least* $1/c_4$ *amount of time, where* c_4 *is a constant independent of* m.

The capacity achieving scheme is as follows.

Capacity Achieving Scheme:

1) At each *odd* time slot, we schedule transmissions from the sources to the relays. We divide the unit square into $g_1(n) = \lfloor \left(\frac{n^{2/3}}{8 \log n} \right)^{\frac{1}{2}} \rfloor^2$ cells. Each cell is a square of area $1/g_1(n)$. We refer to each cell in the odd time slot as a *sending cell*. By Lemma 2.6, each cell can be active for $\frac{1}{c_4}$ amount of time. When a cell is scheduled to be active, each node in the cell broadcasts a new message to all other nodes in the same cell for $\frac{1}{32 c_4 n^{1/3} \log n}$ amount of time (Fig. 2.3). These other nodes then serve as mobile relays for the message. The nodes within the same sending cell coordinate themselves to broadcast sequentially. If any sending cell has more than $32 n^{1/3} \log n$ nodes, we refer to it as a Type-I

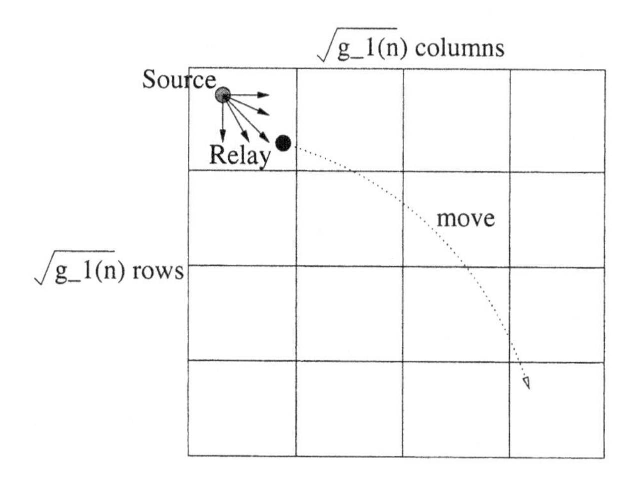

Figure 2.3. Transmission schedule in the odd time slot. $g_1(n) = \lfloor \left(\frac{n^{2/3}}{8 \log n} \right)^{\frac{1}{2}} \rfloor^2$.

error [4]. Unless a Type-I error occurs, each source can broadcast a message of length $\frac{W}{32 c_4 n^{1/3} \log n}$ to all other nodes in the same sending cell.

2) At each *even* time slot, we schedule transmissions from the mobile relays to the destination nodes. Note that the positions of the mobile relays have changed and are now independent of their positions in the previous time slot. We divide the unit square into $g_2(n) = \lfloor (n^{1/3})^{\frac{1}{2}} \rfloor^2$ cells. Each cell is a square of area $1/g_2(n)$. We refer to each cell in the even time slot as the *receiving cell.* For any *receiving cell* $i = 1, ..., g_2(n)$ and any *sending cell* $j = 1, ..., g_1(n)$, pick a node Y_{ij} that is in the *receiving cell* i in the current time slot and that was in the *sending cell* j in the previous time slot. We refer to this node Y_{ij} as the *designated mobile relay in* receiving cell i and *for* sending cell j. If there is no such node Y_{ij} for any i or j, we refer to it as a Type-II error. There may be multiple nodes that can serve as the designated mobile relay for some i, j. In this case we only pick one. Unless a Type-II error occurs, each receiving cell will contain one designated mobile relay from every sending cell. Therefore, each destination node can now find a designated mobile relay that holds the message intended for the destination node and that resides in the same receiving cell (see Fig. 2.4). We then schedule multi-hop transmissions in the following fashion to forward each message from the designated mobile relay to its destination in the same receiving cell. We further divide each receiving cell i into $g_3(n) = \lfloor \left(\frac{n^{2/3}}{4 \log n} \right)^{\frac{1}{2}} \rfloor^2$ *mini-cells* (in $\sqrt{g_3(n)}$ rows and $\sqrt{g_3(n)}$ columns, see Fig. 2.5). Each mini-cell is a square of area $1/(g_2(n) g_3(n))$. By Lemma 2.6,

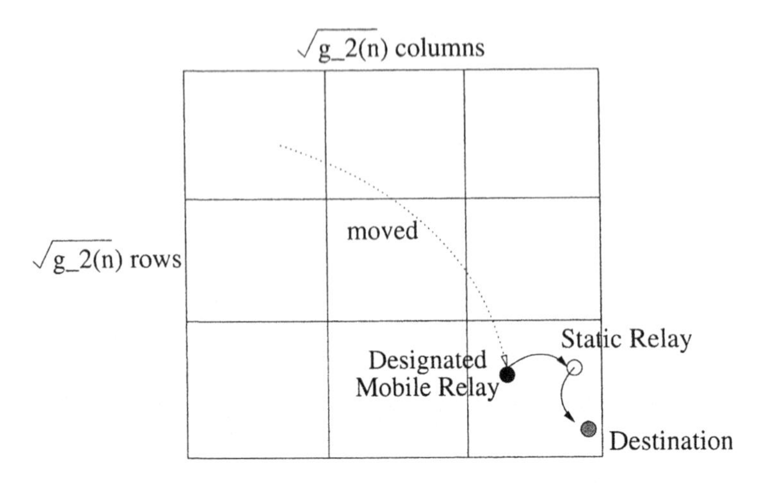

Figure 2.4. Transmission schedule in the even time slot. $g_2(n) = \lfloor \left(n^{1/3}\right)^{\frac{1}{2}} \rfloor^2$.

there exists a scheduling scheme where each mini-cell can be active for $\frac{1}{c_4}$ amount of time. When each mini-cell is active, it forwards a message (or a part of a message) to one other node in the neighboring mini-cell. If the destination of the message is in the neighboring cell, the message is forwarded directly to the destination node. The messages from each designated mobile relay are first forwarded towards neighboring cells along the X-axis, then to their destination nodes along the Y-axis (see Fig. 2.5). In this fashion, a successful schedule will allow each destination node to receive a message of length $\frac{W}{32c_4 n^{1/3}\log n}$ from its respective designated mobile relay residing in the same receiving cell. For details on constructing such a schedule, see Appendix 2.D. If no such schedule exists, we refer to it as a Type-III error. At the end of each even time slot, if there are any packets that cannot be delivered to the destination nodes due to Type-II or Type-III errors, they are dropped.

We can show that, as $n \to \infty$, the probabilities of errors of all types will go to zero. The following proposition thus holds. The proof is available in Appendix 2.D.

PROPOSITION 2.7 *With probability approaching one, as $n \to \infty$, the above scheme allows each source to send a message of length $\frac{W}{32c_4 n^{1/3}\log n}$ to its respective destination node within two time slots.*

Remark: Our scheme uses different cell-partitioning in the odd time slots than that in the even time slots. Note that in previous works [3, 4], the cell structure remains the same over all time slots. Our judicious choice of the cell-structures is the key to our tighter lower bound for the capacity. In particular,

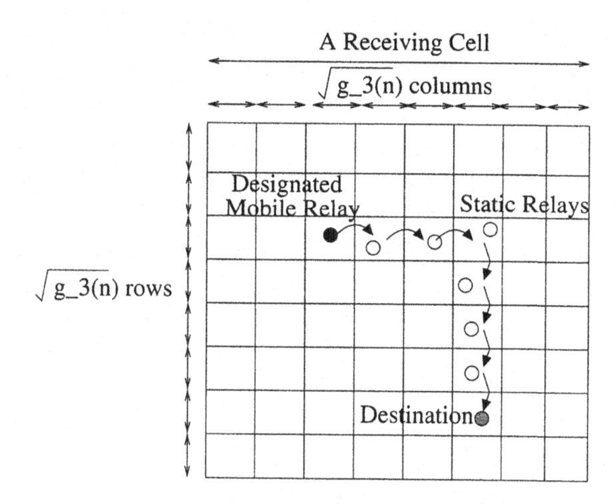

Figure 2.5. Multi-hop transmissions within a receiving cell.

the size of the sending cell is chosen such that the average number of nodes in each cell, $n/g_1(n) = \Theta(n^{1/3}\log n)$, is close to the optimal value of R_b in Section 5.1 (with $d = 0$). The size of the receiving cell is chosen such that its area, $1/g_2(n) = \Theta(n^{1/3})$, is close to the optimal value of l_b^2. Finally, the size of the mini-cell is chosen such that each hop to the neighboring cell is of length $1/\sqrt{g_2(n)g_3(n)} = \Theta(\sqrt{\log n/n})$, which is close to the optimal value of S_b^h.

5.3 The Effect of Queuing

When we defined the delay D_b of each bit b in Section 3, it includes the possible queuing delay at the source node and at the relay nodes. The upper bound on the capacity-delay tradeoff (Proposition 2.5) thus holds regardless of the queuing discipline used in the system, and $\bar{D}$ also includes the queuing delay. We now show how to analyze the queuing delay of the capacity-achieving scheme in Section 5.2. This scheme attempts to deliver one message of length $\frac{W}{32c_4n^{1/3}\log n}$ for each source-destination pair every two time slots. Let p_e be the probability that a message is successfully delivered to the destination at the end of the even time slot. (Note that p_e is the same for all source-destination pairs due to symmetry, and by Proposition 2.7, $p_e \to 1$ as $n \to \infty$.) Assume that, if such delivery is unsuccessful, messages that have not been delivered to the destinations at the end of each even time slot are discarded and have to be retransmitted at the source nodes. Further, assume that packets of length $\frac{W}{32c_4n^{1/3}\log n}$ arrive at each source according to certain stochastic process. Then packets may get enqueued at the source nodes. If we observe the system at the end of each even time slot, the number of packets queued for each source-destination pair will evolve as that of a discrete-time queue with geometric service time distributions [6], and the queues for each source-destination pair can be studied independently. If we know the packet arrival process, we can then compute the queuing delay. For example, if the arrival process is Bernoulli, i.e., one new packet for each source-destination pair arrives at the source every two time slots with probability Λ, then using standard results for discrete time $M/M/1$ queues [6, p82], we can compute the queuing delay as,

$$\mathcal{D} = 2\frac{1-\Lambda}{p_e - \Lambda}.$$

As $n \to \infty$, $p_e \to 1$. Hence,

$$\mathcal{D} \to 2, \text{ as } n \to \infty.$$

On the other hand, if the arrival process is Poisson with rate Λ, then the number of packets arriving at a source-destination pair every two time slots is a Poisson random variable with mean 2Λ. Hence, using results for discrete time $M^{a_n}/M/1$ queues [6, p89], we can compute the queuing delay as

$$\mathcal{D} = 2\frac{1-\Lambda}{p_e - 2\Lambda}.$$

Assume $2\Lambda \leq 1 - \epsilon$, where $0 < \epsilon < 1$. As $n \to \infty$, $p_e \to 1$. Hence,

$$\mathcal{D} \to 2\frac{1-\Lambda}{\epsilon}, \text{ as } n \to \infty.$$

Note that in both cases, the queuing delay is at most a constant multiple of 2 (time slots) provided that ϵ (i.e., the difference between the arrival rate and the capacity) is positive and bounded away from zero as $n \to \infty$. Hence, the capacity-achieving scheme in Section 5.2 can sustain $\Theta(n^{-1/3}/\log n)$ throughput (in bits per time slot) with $O(1)$ *queuing delay*.

5.4 The Capacity Achieving Scheme for Arbitrary Delay Bound

The above scheme can be generalized to arbitrary delay bounds. Let the mean delay be bounded by $\bar{D} = \Theta(n^d)$, $0 \leq d < 1$. We can group every $\lfloor n^d \rfloor + 1$ time slots into a *super-frame*. In each *odd* super-frame, we schedule transmissions from the sources to the relays. We divide the unit square into $\Theta(n^{(2+d)/3}/\log n)$ sending cells of equal area. Within each sending cell, each source broadcasts a new message to all other nodes within the same cell for a duration of $\Theta(\frac{1}{n^{(1-d)/3}\log^2 n})$ every time slot.

In each *even* super-frame, we schedule transmissions from the relays to the destination nodes. We divide the unit square into $\Theta(n^{(1+2d)/3})$ receiving cells of equal area. In every time slot, some mobile relays will have messages intended for some other destination nodes in the same receiving cell. We then schedule multi-hop transmissions to deliver the messages from the mobile relays to the destination nodes in the same receiving cell.

Using similar techniques as the one in [4] and the one in Appendix 2.D, we can show that, with probability approaching one as $n \to \infty$, each source can send $\lfloor n^d \rfloor + 1$ messages of length $\Theta(n^{-(1-d)/3}/\log^2 n)$ to its destination within $2(\lfloor n^d \rfloor + 1)$ time slots. The queuing delay can also be studied in a similar fashion as in Section 5.3. The details are omitted because of space constraints.

6. The Limiting Factors in Existing Schemes

In Section 5, we have shown that choosing the optimal values of the scheduling parameters is the key to achieve the optimal capacity-delay tradeoff. In this section, we will show that deviating from these optimal values will lead to suboptimal capacity-delay tradeoffs. In particular, we will identify the limiting factors in the existing schemes in [3] and [4] by comparing the optimal values of scheduling parameters in Section 5.1 with those used by the existing schemes. Our model in Section 4 can be extended to study the upper bounds on the capacity-delay tradeoff when one imposes additional restrictive assumptions that correspond to these limiting factors. We will see that these new upper bounds are inferior to the capacity-delay tradeoff reported in Sections 4 and 5. The existing schemes of [3] and [4] in fact achieve capacity-delay tradeoffs that are close to the respective upper bounds. These results will give us new insights on which schemes to use under different conditions.

6.1 The Limiting Factor in the Scheme of Neely and Modiano

The scheme by Neely and Modiano [3] divides the unit square into n cells each of area $1/n$. A mobile relay will forward messages to the destination only when they both reside in the same cell. Hence, the distance from the last mobile relay to the destination, l_b, is on average on the order of $O(1/\sqrt{n})$, regardless of the delay constraints. However, we have shown in Section 5.1 that the optimal choice for l_b should be on the order of $\Theta(n^{-(1+2d)/6} \log^{-1/2} n)$, when the mean delay is bounded by $\Theta(n^d)$. The next Proposition shows that the restrictive choice of l_b is indeed the limiting factor of the scheme in [3]. The proof is available in Appendix 2.E

PROPOSITION 2.8 *Let $\bar{D}$ be the mean delay averaged over all bits and all source-destination pairs, and let λ be the throughput of each source-destination pair. If $\bar{D} = O(n^d), 0 \le d < 1$ and $\mathbf{E}[l_b] = O(1/\sqrt{n})$, then for any causal* *scheduling policy,*

$$\lambda \leq O(\frac{\bar{D}}{n} \log^2 n).$$

Remark: The scheme of [3] achieves the above upper bound up to a logarithmic factor.

6.2 The Limiting Factor in the Scheme of Toumpis and Goldsmith

In the scheme by Toumpis and Goldsmith [4], a mobile relay will always use single-hop transmission to forward the messages directly to the destination. That is, the number of hops from the last mobile relay to the destination node, h_b, is always 1. However, we have shown in Section 5.1 that the optimal value of h_b is $\Theta(n^{(1-d)/3}/\log n)$ when the mean delay is bounded by $\Theta(n^d)$. The next Proposition shows that the restriction on h_b is indeed the limiting factor of the scheme in [4]. The proof is available in Appendix 2.F.

PROPOSITION 2.9 *Let $\bar{D}$ be the mean delay averaged over all bits and all source-destination pairs, and let λ be the throughput of each source-destination pair. If $\bar{D} = O(n^d), 0 \leq d < 1$ and $h_b = O(1)$, then for any causal scheduling policy,*

$$\lambda^2 \leq O(\frac{\bar{D}}{n} \log^3 n).$$

Remark: The scheme of [4] achieves the above upper bound up to a logarithmic factor.

Propositions 2.5, 2.8 and 2.9 present three different upper bounds on the capacity-delay tradeoff of mobile wireless networks under different assumptions. Assume that the mean delay is bounded by $n^d, 0 \leq d < 1$. When the capacity is the main concern, Proposition 2.5 shows that the per-node throughput is at most $O(n^{-(1-d)/3} \log n)$. The capacity-achieving scheme reported in Section 5 can achieve close to this upper bound up to a logarithmic factor. However, this capacity-achieving scheme requires sophisticated coordination among the mobile nodes. Hence, it may not be suitable *when simplicity is the main concern.* On the other hand, the scheme of [3] only requires coordination among nodes that are within a cell of area $1/n$. Note that the average number of nodes in such a cell is $\Theta(1)$. Proposition 2.8 then shows that, when coordination among a large number of nodes is prohibited, the scheme of [3] is close to optimal. Similarly, the scheme of [4] only requires single-hop transmissions from the mobile relays to the destinations. Proposition 2.9 shows that, when multi-hop transmissions are undesirable, the scheme of [4] is close to optimal. Therefore, the results reported in this paper present a relatively complete picture of the achievable capacity-delay tradeoffs under different conditions.

An interesting open problem for future work is to investigate whether these insights apply to the capacity-delay tradeoff under mobility models other than the *i.i.d.* model. For example, [7] and [8] have studied the capacity-delay tradeoff under the Brownian Motion mobility model. In these works, the authors also have implicit restrictions on the scheduling policy. In particular, the scheme in [7] uses at most one mobile relay at any time (i.e., $R_b = 1$), and the scheme in [8] schedules a transmission from the mobile relay to the destination only when they are at a distance of $O(1/\sqrt{n})$ away (i.e., $l_b = O(1/\sqrt{n})$). As we have shown in this paper, under the *i.i.d.* mobility model, the optimal capacity-delay tradeoff can only be achieved when R_b, l_b and h_b all vary as functions of the delay exponent d. Putting restrictions on any one of these variables will lead to suboptimal capacity for a given delay constraint. For our future work, we plan to study whether these kind of restrictions will also limit the capacity-delay tradeoff obtained in existing works under other mobility models.

7. Conclusion and Future Work

In this paper, we have studied the fundamental capacity-delay tradeoff in mobile wireless networks under the *i.i.d.* mobility model. Our contributions are three-fold. We have established the upper bound on the optimal capacity-delay tradeoff over all causal scheduling policies. The upper bound not only provides the fundamental limits of capacity and delay, but also helps to identify the optimal values of the key scheduling parameters in order to achieve the optimal capacity-delay tradeoff. Our second contribution is to develop a new scheduling scheme that can achieve a capacity-delay tradeoff that differs from the upper bound only by a logarithmic factor, which also implies that our upper bound is fairly tight. The capacity achievable by our new scheme is larger than that of the existing schemes in [3] and [4]. In particular, when the delay is bounded by a constant, our scheme achieves a per-node capacity of $\Theta(n^{-1/3}/\log n)$. This demonstrates that, under the *i.i.d.* mobility model, mobility increases the capacity even with constant delays. Our third contribution is to use the insight drawn from the upper bound to identify the limiting factors in the existing schemes. These results present a relatively complete picture of the achievable capacity-delay tradeoffs under different considerations.

In this paper, we have assumed an *i.i.d.* mobility model. For future work, we plan to study the optimal capacity-delay tradeoff for mobile wireless networks under other mobility models. Among the properties that we proved in Section 3, we expect that the Tradeoffs II to IV will be relatively invariant to the choice of mobility models, while Tradeoff I is likely to depend on a specific model. Hence, future work will concentrate on how to tailor Tradeoff I for other mobility models. Some immediate extensions to the *i.i.d.* mobility model are possible. For example, at each time slot, each node may independently choose to stay in its old position with probability p, and to move to a new random position with probability $1-p$. This model may approximate scenarios where nodes move at a fast speed and then stay for a relatively long period of time. Tradeoff I will hold for this extension of the *i.i.d.* mobility model, and hence our main results will hold as well. Other mobility models that we plan to investigate are, the Brownian motion mobility model [7, 8], the random waypoint model [8, 9], and the linear mobility model [10], etc.

Other aspects to consider are how the upper bound will be impacted by the use of diversity coding [11], effect of fading [4], and the use of information-theoretic approaches [12, 13].

Appendix: (2.A) Proof of Proposition 2.1

We will need the following lemma on the minimum distance from the mobile relays to the destination at any time slot. Fix a bit b that enters into the system at time slot $t_0(b)$. At each time slot $t \geq t_0(b)$, recall that $r_b(t)$ is the number of mobile relays holding the bit b at the beginning of the time slot. Among these $r_b(t)$ mobile relays, there is one mobile relay whose distance to the destination of bit b is the smallest. Let $\tilde{L}_b(t)$ denote this minimum distance, and let

$$L_b(t) = \max\{\frac{1}{n^2}, \tilde{L}_b(t)\}.$$

It is easy to verify that

$$\tilde{l}_b(t) \geq \tilde{L}_b(t) \geq L_b(t) - \frac{1}{n^2}.$$

LEMMA A.1 *Under the* i.i.d. *mobility model, if* $n \geq 3$, *then*

$$\mathbf{E}\left[\frac{1}{L_b^2(t) r_b(t)} | \mathcal{F}_{t-1}\right] \leq 8\pi \log n \textit{ for all } t \geq t_0(b).$$

Proof: Let $\mathbf{I}_A$ be the indicator function on the set A. By the definition of $L_b(t)$, we have,

$$\begin{aligned}\mathbf{E}\left[\frac{1}{L_b^2(t)}|\mathcal{F}_{t-1}\right] &= \mathbf{E}\left[n^4 \mathbf{I}_{\{\tilde{L}_b(t) \leq \frac{1}{n^2}\}}|\mathcal{F}_{t-1}\right] \\ &\quad + \mathbf{E}\left[\frac{1}{\tilde{L}_b^2(t)}\mathbf{I}_{\{\tilde{L}_b(t) > \frac{1}{n^2}\}}|\mathcal{F}_{t-1}\right].\end{aligned}$$

Since the nodes move on a unit square, $\tilde{L}_b(t) \leq \sqrt{2}$. Hence,

$$\begin{aligned}&\mathbf{E}\left[\frac{1}{\tilde{L}_b^2(t)}\mathbf{I}_{\{\tilde{L}_b(t) > \frac{1}{n^2}\}}|\mathcal{F}_{t-1}\right] \\ =& \int_{\frac{1}{n^2}}^{\sqrt{2}} \frac{1}{u^2} d\mathbf{P}[\tilde{L}_b(t) \leq u|\mathcal{F}_{t-1}] \\ =& \frac{1}{u^2}\mathbf{P}[\tilde{L}_b(t) \leq u|\mathcal{F}_{t-1}]|_{\frac{1}{n^2}}^{\sqrt{2}} - \int_{\frac{1}{n^2}}^{\sqrt{2}} \mathbf{P}[\tilde{L}_b(t) \leq u|\mathcal{F}_{t-1}] d\frac{1}{u^2} \\ =& \frac{1}{2} - n^4 \mathbf{P}[\tilde{L}_b(t) \leq \frac{1}{n^2}|\mathcal{F}_{t-1}] + \int_{\frac{1}{n^2}}^{\sqrt{2}} \frac{2}{u^3}\mathbf{P}[\tilde{L}_b(t) \leq u|\mathcal{F}_{t-1}] du.\end{aligned}$$

Hence,

$$\mathbf{E}\left[\frac{1}{L_b^2(t)}|\mathcal{F}_{t-1}\right] = \frac{1}{2} + \int_{\frac{1}{n^2}}^{\sqrt{2}} \frac{2}{u^3}\mathbf{P}[\tilde{L}_b(t) \le u|\mathcal{F}_{t-1}]du.$$

Let ρ_b be the distance from any one mobile node to the destination of the bit b. Then, due to the *i.i.d.* mobility model, we have,

$$\mathbf{P}[\rho_b \le u|\mathcal{F}_{t-1}] \le \pi u^2,$$

and,

$$\begin{aligned}\mathbf{P}[\tilde{L}_b(t) \le u|\mathcal{F}_{t-1}] &\le 1-(1-\pi u^2)^{r_b(t)}\\ &\le \pi r_b(t)u^2.\end{aligned}$$

Therefore,

$$\begin{aligned}\mathbf{E}\left[\frac{1}{L_b^2(t)}|\mathcal{F}_{t-1}\right] &= \frac{1}{2} + \int_{\frac{1}{n^2}}^{\sqrt{2}} \frac{2}{u^3}\mathbf{P}[\tilde{L}_b(t) \le u|\mathcal{F}_{t-1}]du\\ &\le \frac{1}{2} + \int_{\frac{1}{n^2}}^{\sqrt{2}} \pi r_b(t)\frac{2}{u}du\\ &= \frac{1}{2} + 2\pi r_b(t)\log u|_{\frac{1}{n^2}}^{\sqrt{2}}\\ &= \frac{1}{2} + 2\pi r_b(t)(\log\sqrt{2} + 2\log n)\\ &\le 8\pi r_b(t)\log n,\end{aligned}$$

when $n \ge 3$. Finally, since $r_b(t)$ is $\mathcal{F}_{t-1}$-measurable, we have

$$\begin{aligned}\mathbf{E}\left[\frac{1}{L_b^2(t)r_b(t)}|\mathcal{F}_{t-1}\right] &= \frac{1}{r_b(t)}\mathbf{E}\left[\frac{1}{L_b^2(t)}|\mathcal{F}_{t-1}\right]\\ &\le 8\pi\log n.\end{aligned}$$

Q.E.D.

Proof of Proposition 2.1 : Let

$$V_t = 8\pi\log n\,[t - t_0(b)] - \sum_{s=t_0(b)+1}^{t} \frac{1}{L_b^2(t)r_b(t)}\mathbf{I}_{\{C_b(t)=1\}},$$

Then for all $t \ge t_0(b)$, V_t is also $\mathcal{F}_t$-measurable and $V_{t_0(b)} = 0$. By Lemma A.1, we have

$$\begin{aligned}&\mathbf{E}[V_t - V_{t-1}|\mathcal{F}_{t-1}]\\ =\ &8\pi\log n - \mathbf{E}\left[\frac{1}{L_b^2(t)r_b(t)}\mathbf{I}_{\{C_b(t)=1\}}|\mathcal{F}_{t-1}\right]\\ \ge\ &8\pi\log n - \mathbf{E}\left[\frac{1}{L_b^2(t)r_b(t)}|\mathcal{F}_{t-1}\right]\\ \ge\ &0.\end{aligned}$$

Hence,

$$\mathbf{E}[V_t|\mathcal{F}_{t-1}] \ge V_{t-1},$$

i.e., V_t is a sub-martingale. Recall that $s_b = \min\{t : t \geq t_0(b) \text{ and } C_b(t) = 1\}$. Since s_b is a stopping time, by appropriately invoking the Optional Stopping Theorem [5, p249, Theorem 4.1], we have,

$$\mathbf{E}[V_{s_b}] \geq 0.$$

Hence,

$$8\pi \log n \mathbf{E}[D_b] \geq \mathbf{E}\left[\frac{1}{L_b^2(s_b)R_b}\right].$$

Using Hőlder's Inequality [5, p15]

$$\mathbf{E}^2[\frac{1}{L_b(s_b)}] \leq \mathbf{E}[R_b]\mathbf{E}[\frac{1}{L_b^2(s_b)R_b}],$$

we have,

$$\begin{aligned} 8\pi \log n \mathbf{E}[D_b] &\geq \mathbf{E}^2[\frac{1}{L_b(s_b)}]\frac{1}{\mathbf{E}[R_b]} \\ &\geq \frac{1}{\mathbf{E}^2[L_b(s_b)]\mathbf{E}[R_b]}. \end{aligned}$$

Finally, by definition,

$$l_b = \tilde{l}_b(s_b) \geq L_b(s_b) - \frac{1}{n^2},$$

therefore,

$$8\pi \log n \mathbf{E}[D_b] \geq \frac{1}{(\mathbf{E}[l_b] + \frac{1}{n^2})^2 \mathbf{E}[R_b]}.$$

Q.E.D.

Appendix: (2.B) Proof of Proposition 2.3

The next Lemma will be used frequently in the proof of Propositions 2.3 and 2.7. Consider an experiment where we randomly throw n balls into $m \leq n$ urns. The probability that each ball j enters urn i is $\frac{p}{m}$ and is independent of the position of other balls. Thus, $p \leq 1$ is the *success probability* that the ball is thrown into *any* one of the urns. Let $B_i, i = 1, ...m$ be the number of balls in urn i after n balls are thrown. It is obvious that $\mathbf{E}[B_i] = \frac{np}{m}$. The following Lemma shows that, when n is large, with high probability all B_i will concentrate on its mean.

LEMMA B.1 *As $n \to \infty$,*

1) If $\frac{np}{m} = c$, where c is a positive constant, then

$$\mathbf{P}[B_i \geq \frac{np}{m} \log n \textit{ for any } i] = O(\frac{1}{n}).$$

2) If $\frac{np}{m} \geq c \log n$ and $c \geq 8$, then

$$\mathbf{P}[B_i \geq 2\frac{np}{m} \textit{ for any } i] \leq \frac{1}{n}.$$

3) If $\frac{np}{m} \geq cn^\alpha$, where $c > 0$ and $\alpha > 0$, then

$$\mathbf{P}[B_i \geq 2\frac{np}{m} \textit{ for any } i] = O(\frac{1}{n}).$$

4) If $\frac{np}{m} \geq c \log n$ *and* $c \geq 4$, *then*

$$\mathbf{P}[B_i = 0 \textit{ for any } i] = O(\frac{1}{n}).$$

5) If $\frac{np}{m} \geq c \log n$ *and* $c \geq 16$, *then*

$$\mathbf{P}[B_i \geq 2\frac{np}{m} \textit{ for any } i] \leq \frac{1}{n^3}.$$

Proof: By known results on the characteristic function of Bernoulli random variables, we have, for any $\theta > 0$,

$$\begin{aligned}
\mathbf{E}[e^{\theta B_i}] &= \left[e^{\theta}\frac{p}{m} + (1 - \frac{p}{m})\right]^n \\
&= \left[1 + (e^{\theta} - 1)\frac{p}{m}\right]^n \\
&\leq \exp\left[\frac{np}{m}(e^{\theta} - 1)\right] \text{ for all urn } i,
\end{aligned}$$

where in the last step we have used the inequality that

$$(1 + x)^{\frac{1}{x}} \leq e \text{ for } x > 0.$$

Using the Markov Inequality [5, p15], for any $y > 0$,

$$\begin{aligned}
\mathbf{P}[B_i \geq y] &\leq \frac{\mathbf{E}[e^{\theta B_i}]}{e^{\theta y}} \\
&\leq \exp\left[\frac{np}{m}(e^{\theta} - 1) - \theta y\right].
\end{aligned}$$

Hence, by the union bound

$$\mathbf{P}[B_i \geq y \text{ for any } i] \leq n \exp\left[\frac{np}{m}(e^{\theta} - 1) - \theta y\right].$$

To prove part 1, let $\frac{np}{m} = c$ and $y = \frac{np}{m} \log n$, then

$$\begin{aligned}
\mathbf{P}[B_i \geq y \text{ for any } i] &\leq n \exp\left[\frac{np}{m}(e^{\theta} - 1) - \theta y\right] \\
&= n \exp\left[c(e^{\theta} - 1) - \theta c \log n\right].
\end{aligned}$$

Let $\theta = 2/c$, then

$$\begin{aligned}
\mathbf{P}[B_i \geq y \text{ for any } i] &\leq n\frac{1}{n^2} \exp\left[c(e^{2/c} - 1)\right] \\
&= O(\frac{1}{n}).
\end{aligned}$$

To prove part 2, let $y = 2\frac{np}{m}$, hence

$$\begin{aligned}
\mathbf{P}[B_i \geq y \text{ for any } i] &\leq n \exp\left[\frac{np}{m}(e^{\theta} - 1) - \theta y\right] \\
&= n \exp\left[\frac{np}{m}(e^{\theta} - 1 - 2\theta)\right]. \qquad (2.B.1)
\end{aligned}$$

Let $\theta = \log 2$, then

$$e^{\theta} - 1 - 2\theta = -(2\log 2 - 1) = -0.386 \le -\frac{1}{4}.$$

Hence, when $\frac{np}{m} \ge c \log n$ and $c \ge 8$, we have

$$\begin{aligned} \mathbf{P}[B_i \ge y \text{ for any } i] &\le n \exp\left[-\frac{c}{4}\log n\right] \\ &\le n\frac{1}{n^2} = \frac{1}{n}. \end{aligned}$$

To prove part 3, substituting $\theta = \log 2$ and $\frac{np}{m} \ge cn^{\alpha}$ into (2.B.1), we have,

$$\mathbf{P}[B_i \ge y \text{ for any } i] \le n \exp\left[-\frac{c}{4}n^{\alpha}\right] \le O(\frac{1}{n}).$$

To prove part 4, note that for any i,

$$\begin{aligned} \mathbf{P}[B_i = 0] &= \left[1 - \frac{p}{m}\right]^n \\ &\le \left[1 - \frac{4\log n}{n}\right]^n \\ &= \left[1 - \frac{4\log n}{n}\right]^{\frac{n}{4\log n}4\log n}. \end{aligned}$$

Since $\lim_{x \to 0}(1-x)^{1/x} = 1/e$, we have

$$\left[1 - \frac{4\log n}{n}\right]^{\frac{n}{4\log n}} \le \frac{1}{\sqrt{e}} \text{ for large } n.$$

Hence, for large n,

$$\mathbf{P}[B_i = 0] \le \left[\frac{1}{\sqrt{e}}\right]^{4\log n} = \frac{1}{n^2}.$$

Therefore,

$$\mathbf{P}[B_i = 0 \text{ for any } i\,] \le n\frac{1}{n^2} = O(\frac{1}{n}).$$

Part 5 can be shown analogously as Part 2. *Q.E.D.*

Proof of Proposition 2.3 : At each time slot t, an opportunistic broadcast scheme has to determine how to duplicate each bit b to a larger number of mobile nodes. Some of the mobile nodes that already have the bit b have to be selected to transmit the bit b, and some of the other mobile nodes have to be selected to receive the bit.

Let $v_b(t,i)$ be the distance from a node i that is chosen to transmit the bit b at time slot t, to the furthest node that is chosen to receive the bit ($v_b(t,i) = 0$ if node i is not chosen to transmit the bit or if the bit b has cleared the system). Let $u_b(t,i)$ be the number of mobile nodes that are chosen to receive the bit b from node i and that do not have the bit b prior to time slot t ($u_b(t,i) = 0$ if $v_b(t,i) = 0$). Then $u_b(t,i)$ is bounded from above by the number of nodes covered by a disk of radius $v_b(t,i)$ centered at node i. It is easy to verify that

$$R_b - 1 = \sum_{t=1}^{T}\sum_{i=1}^{n} u_b(t,i).$$

Fix a time slot t. We next bound the number of nodes that is covered by each disk of radius $v_b(t,i)$ centered at node i. We divide the unit square into $g_4(n) = \lfloor \left(\frac{n}{16\log n}\right)^{\frac{1}{2}} \rfloor^2$ cells (in $\sqrt{g_4(n)}$ rows and $\sqrt{g_4(n)}$ columns). Each cell is a square of area $1/g_4(n)$. Let B_i be the number of nodes in cell i, $i = 1, ..., g_4(n)$. Then $\mathbf{E}[B_i] = \frac{n}{g_4(n)}$. When n is large, we have

$$16\log n \leq \frac{n}{g_4(n)} \leq 32\log n.$$

Let $\mathcal{A}$ be the event that

$$B_i \leq \frac{2n}{g_4(n)} \text{ for all } i = 1, ..., g_4(n).$$

By part 5 of Lemma B.1, $\mathbf{P}[\mathcal{A}^c] \leq 1/n^3$, Now consider each disk of radius $v_b(t,i)$. We need at most

$$\left[2v_b(t,i)\sqrt{g_4(n)} + 2\right]^2$$

cells to completely cover the disk. Hence, if event $\mathcal{A}$ occurs, the number of nodes in the disk of radius $v_b(t,i)$ will be bounded from above by

$$\begin{aligned} & \left[2v_b(t,i)\sqrt{g_4(n)} + 2\right]^2 \frac{2n}{g_4(n)} \\ \leq \; & 16nv_b^2(t,i) + \frac{16n}{g_4(n)} \\ \leq \; & 16nv_b^2(t,i) + 512\log n \end{aligned}$$

Note that the above relationship holds for all b and i. Let $c_7 = 16/\pi$, and $c_8 = 512$. Since $u_b(t,i)$ is no greater than the number of nodes covered by the disk of radius $v_b(t,i)$, we have,

$$\mathbf{P}\left[\frac{u_b(t,i)}{n} > c_7\pi v_b^2(t,i) + c_8\frac{\log n}{n} \text{ for any } b, i\right] \leq \mathbf{P}[\mathcal{A}^c] \leq \frac{1}{n^3}.$$

Fix a bit b. Let $\mathcal{B}$ be the event that

$$\frac{u_b(t,i)}{n} \leq c_7\pi v_b^2(t,i) + c_8\frac{\log n}{n} \text{ for all } i \text{ and } t = t_0(b), ..., t_0(b) + c_2n^2.$$

Then,

$$\mathbf{P}[\mathcal{B}^c] \leq \frac{1}{n^3}c_2n^2 = \frac{c_2}{n}.$$

Since $u_b(t,i) \leq n$, we have

$$\begin{aligned} \mathbf{E}\left[\frac{u_b(t,i)}{n}\right] &= \mathbf{E}\left[\frac{u_b(t,i)}{n}\mathbf{I}_{\{\mathcal{B}\}}\right] + \mathbf{E}\left[\frac{u_b(t,i)}{n}\mathbf{I}_{\{\mathcal{B}^c\}}\right] \\ &\leq \mathbf{E}\left[c_7\pi v_b^2(t,i) + c_8\frac{\log n}{n}\right] + \mathbf{P}[\mathcal{B}^c] \\ &\leq c_7\pi\mathbf{E}[v_b^2(t,i)] + c_8\frac{\log n}{n} + \frac{c_2}{n} \\ &\leq c_7\pi\mathbf{E}[v_b^2(t,i)] + (c_8+1)\frac{\log n}{n} \end{aligned}$$

when $n \geq \max\{N_0, \exp(c_2)\}$,

We now use the idea in Section 3 that disks of radius $\frac{\Delta}{2}$ times the transmission range centered at the transmitter are disjoint from each other. For each unicast transmission (i.e., the transmission over each hop S_b^h), the transmission range is just S_b^h. For broadcast, the transmission range is the distance from the transmitter to the furthest node that can successfully receive the bit, i.e. $v_b(t,i)$. By counting the area covered by all the disks, we have

$$\sum_{b=1}^{\lambda nT}\sum_{t=1}^{T}\sum_{i=1}^{n}\pi\frac{\Delta^2}{4}v_b^2(t,i)+\sum_{b=1}^{\lambda nT}\sum_{h=1}^{h_b}\frac{\pi\Delta^2}{4}(S_b^h)^2\leq WT. \tag{2.B.2}$$

Since there are at most n nodes that can serve as transmitters at any time, we have

$$\sum_{b=1}^{\lambda nT}\sum_{t=1}^{T}\sum_{i=1}^{n}\mathbf{I}_{\{v_b(t,i)>0\}}\leq WTn.$$

Hence,

$$\begin{aligned}\sum_{b=1}^{\lambda nT}\frac{\mathbf{E}[R_b]-1}{n} &= \mathbf{E}\left[\sum_{b=1}^{\lambda nT}\sum_{t=1}^{T}\sum_{i=1}^{n}\frac{u_b(t,i)}{n}\right]\\ &\leq c_7\pi\mathbf{E}\left[\sum_{b=1}^{\lambda nT}\sum_{t=1}^{T}\sum_{i=1}^{n}v_b^2(t,i)\right]\\ &\quad+(c_8+1)\mathbf{E}\left[\sum_{b=1}^{\lambda nT}\sum_{t=1}^{T}\sum_{i=1}^{n}\frac{\log n}{n}\mathbf{I}_{\{v_b(t,i)>0\}}\right]\\ &\leq c_7\pi\mathbf{E}\left[\sum_{b=1}^{\lambda nT}\sum_{t=1}^{T}\sum_{i=1}^{n}v_b^2(t,i)\right]+(c_8+1)WT\log n. \end{aligned}\tag{2.B.3}$$

Substituting (2.B.3) into (2.B.2), we have

$$\begin{aligned}&\sum_{b=1}^{\lambda nT}\frac{\Delta^2}{4}\frac{\mathbf{E}[R_b]-1}{n}+\sum_{b=1}^{\lambda nT}\sum_{h=1}^{h_b}\frac{\pi\Delta^2}{4}\mathbf{E}[(S_b^h)^2]\\ \leq&\left\{c_7\mathbf{E}\left[\sum_{b=1}^{\lambda nT}\sum_{t=1}^{T}\sum_{i=1}^{n}\pi\frac{\Delta^2}{4}v_b^2(t,i)\right]+\sum_{b=1}^{\lambda nT}\sum_{h=1}^{h_b}\frac{\pi\Delta^2}{4}\mathbf{E}[(S_b^h)^2]\right\}\\ &+\frac{(c_8+1)\Delta^2}{4}WT\log n\\ \leq& c_7WT+\frac{(c_8+1)\Delta^2}{4}WT\log n\\ \leq&\frac{(c_8+2)\Delta^2}{4}WT\log n\end{aligned}$$

when $n\geq\max\{N_0,\exp(c_2),\exp(\frac{4c_7}{\Delta^2})\}$. *Q.E.D.*

Appendix: (2.C) Proof of Lemma 2.6

We can group all cells into $c_4=\lfloor 2\Delta+6\rfloor^2$ lattices. Each lattice consists of nodes that are $\lfloor 2\Delta+6\rfloor/m$ apart along the X-axis or the Y-axis (see Fig.2.2). The cells of each lattice can be active at the same time since

- the transmission range from a node to a neighboring node is at most $2/m$, and
- any interfering transmitters are at least $(2\Delta+2)/m$ distance away from the receiver.

We can schedule the c_4 lattices in a round-robin fashion and each lattice is active for $1/c_4$ amount of time.

Appendix: (2.D) Proof of Proposition 2.7

We only need to show that the probabilities of errors of all types will go to zero as $n \to \infty$. Let x_j be the number of nodes in sending cell $j = 1, ..., g_1(n)$. Let $p_{\rm I}$ be the probability of the Type-I error, i.e., $x_j \geq 32n^{1/3} \log n$ for any j. Equivalently, we can consider the experiment that we throw n balls into $g_1(n)$ urns with success probability $p = 1$. It is easy to show that

$$\frac{n^{2/3}}{16 \log n} \leq g_1(n) \leq \frac{n^{2/3}}{8 \log n} \text{ when } n \text{ is large.}$$

The average number of nodes in each sending cell is then

$$\frac{n}{g_1(n)} = \Theta(n^{1/3} \log n).$$

Hence, by part 3 of Lemma B.1,

$$\begin{aligned} p_{\rm I} &= \mathbf{P}[x_j \geq 32n^{1/3} \log n \text{ for any } j] \\ &\leq \mathbf{P}[x_j \geq \frac{2n}{g_1(n)} \text{ for any } j] \\ &= O(1/n). \end{aligned}$$

Let $p_{\rm II}$ be the probability of the Type-II error. For each receiving cell i and sending cell j, let y_{ij} be the number of nodes that are in the receiving cell i in the even time slot and in the sending cell j in the previous odd time slot. Equivalently, we can consider the experiment that we throw n balls to $g_1(n)g_2(n)$ urns with success probability $p = 1$. Since

$$\mathbf{E}[y_{ij}] = \frac{n}{g_1(n)g_2(n)} \geq 8 \log n,$$

by part 4 of Lemma B.1,

$$p_{\rm II} = \mathbf{P}[y_{ij} = 0 \text{ for any } i, j] = O(\frac{1}{n}).$$

Hence, with probability $(1 - p_{\rm I} - p_{\rm II})$ approaching one as $n \to \infty$, each destination node can now find a designated mobile relay that holds the message intended for the destination node and that resides in the same receiving cell. Next we study $p_{\rm III}$, the probability of the Type-III error. We need to specify how to schedule the hop-by-hop transmissions from the designated mobile relays to the destination nodes within each receiving cell. We divide each receiving cell i into $g_3(n) = \lfloor \left(\frac{n^{2/3}}{4 \log n} \right)^{\frac{1}{2}} \rfloor^2$ *mini-cells* (in $\sqrt{g_3(n)}$ rows and $\sqrt{g_3(n)}$ columns, see Fig. 2.5). Each mini-cell is a square of area $1/(g_2(n)g_3(n))$. By Lemma 2.6, there exists a scheduling scheme where each mini-cell can be active for $\frac{1}{c_4}$ amount of time. When each mini-cell is active, it forwards a message (or a part of a message) to one other node in the neighboring mini-cell. If the destination of the message is in the neighboring cell, the message is forwarded directly to the destination node. The messages from each designated mobile relay are first forwarded towards neighboring cells along the X-axis, then to their destination nodes along the Y-axis (see Fig. 2.5).

Note that there are totally n messages of length $\frac{W}{32c_4 n^{1/3} \log n}$ that need to be scheduled. The above scheduling scheme can successfully forward all messages from the designated mobile relays to the destinations provided that:

- Each mini-cell contains at least one node. Hence, each node can always find some node in the neighboring cell to serve as static relays.

- The number of messages that go through any mini-cell is bounded by $32n^{1/3}\log n$. Because each message is of length $\frac{W}{32c_4 n^{1/3}\log n}$, each mini-cell thus only needs to be active for at most $\frac{1}{c_4}$ amount of time, which is always possible by Lemma 2.6[5].

In order to show that p_{III} goes to zero as $n \to \infty$, we only need to show that both of the above conditions will hold with probability approaching one. First note that the average number of nodes in a mini-cell is

$$\frac{n}{g_2(n)g_3(n)} \geq 4\log n.$$

Let $p_{\text{III}}^{\text{a}}$ be the probability that any of the $g_2(n)g_3(n)$ mini-cells are empty. Equivalently, we can consider the experiment that we throw n balls into $g_2(n)g_3(n)$ urns with success probability $p = 1$. Then, by part 4 of Lemma B.1,

$$p_{\text{III}}^{\text{a}} = O(1/n).$$

Next we group the nodes in each receiving cell by the positions of their corresponding source nodes in the previous time slot. Let $\mathcal{Y}_{ij}$ be the set of nodes in the receiving cell i that are the destination nodes for some source nodes in the sending cell j (in the previous time slot). Let $p_{\text{III}}^{\text{b}}$ be the probability that any set $\mathcal{Y}_{ij}, i = 1, ..., g_2(n), j = 1, ..., g_1(n)$, has more than $32\log n$ nodes. Equivalently, we can consider the experiment that we throw n balls into $m = g_1(n)g_2(n)$ urns with success probability $p = 1$. Since

$$16\log n \geq \frac{np}{m} = \frac{n}{g_1(n)g_2(n)} \geq 8\log n \text{ for large } n,$$

by part 2 of Lemma B.1,

$$\begin{aligned} p_{\text{III}}^{\text{b}} &= \mathbf{P}[|\mathcal{Y}_{ij}| \geq 32\log n, \text{ for any } i,j] \\ &\leq \mathbf{P}[|\mathcal{Y}_{ij}| \geq 2\frac{np}{m} \text{ for any } i,j] \\ &= O(\frac{1}{n}). \end{aligned}$$

Hence, with high probability, each designated mobile relay will serve no more than $32\log n$ destination nodes in the same receiving cell. As presented earlier, the message will first be forwarded along the X-axis, then along the Y-axis. We next bound the number of messages that go through any mini-cell along the X-axis. Within a given receiving cell i, fix any mini-cell $k = 1, ..., g_3(n)$. Let $Z_{i,k}^x$ be the number of designated mobile relays in receiving cell i that reside at the same *row* with the mini-cell k. Note that there are at most $g_1(n)$ designated mobile relays and $\sqrt{g_3(n)}$ rows of mini-cells in a given receiving cell i. Let $p_{\text{III}}^{\text{c}}(i)$ be the probability that $Z_{i,k}^x \geq \frac{1}{2}\sqrt{\frac{2n^{2/3}}{\log n}}$ for any mini-cell k in a given receiving cell i. Equivalently, we can consider the experiment that we throw $g_1(n)$ balls into $\sqrt{g_3(n)}$ urns with success probability $p = 1$. The average number of balls per urn is $g_1(n)/\sqrt{g_3(n)}$, and

$$\frac{1}{4}\sqrt{\frac{2n^{2/3}}{\log n}} \geq \frac{g_1(n)}{\sqrt{g_3(n)}} \geq \frac{1}{8}\sqrt{\frac{n^{2/3}}{\log n}} \text{ for large } n.$$

By part 3 of Lemma B.1,

$$p_{\text{III}}^{\text{c}}(i) = \mathbf{P}[Z_{i,k}^x \geq \frac{1}{2}\sqrt{\frac{2n^{2/3}}{\log n}} \text{ for any } k]$$

$$\begin{aligned} &\le \mathbf{P}[Z^x_{i,k} \ge 2\frac{g_1(n)}{\sqrt{g_3(n)}} \text{ for any } k] \\ &= O(\frac{1}{g_1(n)}) = O(n^{-\frac{2}{3}} \log n). \end{aligned}$$

Let $p^{\text{c}}_{\text{III}}$ be the probability that $Z^x_{i,k} \ge \frac{1}{2}\sqrt{\frac{2n^{2/3}}{\log n}}$ for any mini-cell in *any* receiving cell. Since there are $\Theta(n^{1/3})$ receiving cells, by the union bound,

$$\begin{aligned} p^{\text{c}}_{\text{III}} &\le \sum_{i=1}^{g_2(n)} p^{\text{c}}_{\text{III}}(i) \\ &\le n^{\frac{1}{3}} O(n^{-\frac{2}{3}} \log n) \\ &= O(n^{-\frac{1}{3}} \log n). \end{aligned}$$

Therefore, with high probability, there will be at most $\frac{1}{2}\sqrt{\frac{2n^{2/3}}{\log n}}$ designated mobile relays that reside at the same row as any mini-cell, and each of them is the origin of at most $32 \log n$ messages. Hence, with probability approaching one as $n \to \infty$, the number of messages that have to go through any mini-cell along the X-axis is less than

$$16\sqrt{2n^{2/3} \log n}.$$

Similarly, let $Z^y_{i,k}$ be the number of nodes in the receiving cell i that reside at the same *column* as the mini-cell k. Let $p^{\text{d}}_{\text{III}}(i)$ be the probability that $Z^y_{i,k} \ge 8n^{1/3}\sqrt{2\log n}$ for any mini-cell k in a given receiving cell i. Equivalently, we can consider the experiment that we throw n balls into $\sqrt{g_3(n)}$ urns with success probability $p = 1/g_2(n)$. This experiment is independent of the previous one, because the X-coordinates of the nodes are independent of their Y-coordinates. The average number of balls per urn is $n/(g_2(n)\sqrt{g_3(n)})$, and

$$4n^{1/3}\sqrt{2\log n} \ge \frac{n}{g_2(n)\sqrt{g_3(n)}} \ge 2n^{1/3}\sqrt{\log n} \text{ for large } n.$$

Hence, by part 3 of Lemma B.1,

$$\begin{aligned} p^{\text{d}}_{\text{III}}(i) &= \mathbf{P}[Z^y_{i,k} \ge 8n^{1/3}\sqrt{2\log n} \text{ for any } k\,] \\ &\le \mathbf{P}[Z^y_{i,k} \ge 2\frac{n}{g_2(n)\sqrt{g_3(n)}} \text{ for any } k\,] \\ &= O(\frac{1}{n}). \end{aligned}$$

Let $p^{\text{d}}_{\text{III}}$ be the probability that $Z^y_{i,k} \ge 8n^{1/3}\sqrt{2\log n}$ for any mini-cell in *any* receiving cell. By the union bound,

$$\begin{aligned} p^{\text{d}}_{\text{III}} &\le \sum_{i=1}^{g_2(n)} p^{\text{d}}_{\text{III}}(i) \\ &= n^{1/3} O(\frac{1}{n}) = O(n^{-2/3}). \end{aligned}$$

Therefore, with probability approaching one as $n \to \infty$, the number of messages that have to go through any mini-cell along the Y-axis is less than

$$8n^{1/3}\sqrt{2\log n}.$$

Combining all of the above results, with probability no less than

$$1 - p_{\text{III}} \geq 1 - (p_{\text{III}}^{\text{a}} + p_{\text{III}}^{\text{b}} + p_{\text{III}}^{\text{c}} + p_{\text{III}}^{\text{d}}) = 1 - O(n^{-1/3} \log n),$$

the number of messages that have to go through any mini-cell $k = 1, ...g_3(n)$ in any receiving cell i is less than

$$24n^{1/3}\sqrt{2\log n} \leq 32n^{1/3} \log n \text{ for large } n.$$

Proposition 2.7 then follows. *Q.E.D.*

Appendix: (2.E) Proof of Proposition 2.8

We start from inequality (2.17). Since $\mathbf{E}[l_b] \leq \sqrt{c_5}n^{-1/2}$ for some positive constant c_t, we have,

$$\begin{aligned}
\frac{1}{c_1 n \log n} \frac{(\lambda nT)^3}{\bar{D}(2\sum_{b=1}^{\lambda nT} \mathbf{E}[l_b])^2} &\geq \frac{1}{4c_1 n \log n} \frac{(\lambda nT)^3}{\bar{D}(\lambda nT)^2 c_5 n^{-1}} \\
&= \frac{1}{4c_1 c_5 \log n} \frac{\lambda nT}{\bar{D}},
\end{aligned}$$

and,

$$\begin{aligned}
\frac{2\pi}{WTn}(\sum_{b=1}^{\lambda nT} \mathbf{E}[l_b])^2 &\leq \frac{2\pi}{WTn}(\lambda nT)^2 c_5 n^{-1} \\
&= \frac{2\pi c_5}{W}\lambda^2 T.
\end{aligned}$$

Substitute the above two inequalities into (2.17). Note that when $\bar{D} = o(n)$, the first term of (2.17) dominates the rest for large n. Hence

$$\begin{aligned}
\frac{4c_3}{\Delta^2} WT \log n &\geq \frac{1}{2} \frac{1}{c_1 n \log n} \frac{(\lambda nT)^3}{\bar{D}(2\sum_{b=1}^{\lambda nT} \mathbf{E}[l_b])^2} \\
&\geq \frac{1}{8c_1 c_5 \log n} \frac{\lambda nT}{\bar{D}}.
\end{aligned}$$

We can then solve for λ,

$$\lambda \leq \frac{\bar{D} \log^2 n}{n} \frac{32c_1 c_3 c_5 W}{\Delta^2}.$$

Q.E.D.

Appendix: (2.F) Proof of Proposition 2.9

Since $h_b \leq c_6$ for some positive number c_6, we have,

$$\begin{aligned}
\left(\sum_{b=1}^{\lambda nT}\sum_{h=1}^{h_b} S_b^h\right)^2 &\leq \left(\sum_{b=1}^{\lambda nT}\sum_{h=1}^{h_b} 1\right)\left(\sum_{b=1}^{\lambda nT}\sum_{h=1}^{h_b} (S_b^h)^2\right) \\
&\leq \lambda nTc_6 \sum_{b=1}^{\lambda nT}\sum_{h=1}^{h_b} (S_b^h)^2.
\end{aligned}$$

Hence,

$$\begin{aligned}
\mathbf{E}[\sum_{b=1}^{\lambda nT}\sum_{h=1}^{h_b}(S_b^h)^2] &\geq \frac{1}{c_6\lambda nT}\mathbf{E}[\left(\sum_{b=1}^{\lambda nT}\sum_{h=1}^{h_b}S_b^h\right)^2] \\
&\geq \frac{1}{c_6\lambda nT}\left(\mathbf{E}[\sum_{b=1}^{\lambda nT}\sum_{h=1}^{h_b}S_b^h]\right)^2 \\
&\geq \frac{1}{c_6\lambda nT}\left(\sum_{b=1}^{\lambda nT}\mathbf{E}[l_b]\right)^2. \qquad (2.F.1)
\end{aligned}$$

Substitute (2.15) and (2.F.1) into (2.7), we have,

$$\begin{aligned}
\frac{4c_3}{\Delta^2}WT\log n &\geq \sum_{b=1}^{\lambda nT}\frac{\mathbf{E}[R_b]-1}{n} + \pi\mathbf{E}[\sum_{b=1}^{\lambda nT}\sum_{h=1}^{h_b}(S_b^h)^2] \\
&\geq \frac{1}{c_1 n\log n}\frac{(\sum_{b=1}^{\lambda nT}1)^3}{\bar{D}\left(\sum_{b=1}^{\lambda nT}(\mathbf{E}[l_b]+\frac{1}{n^2})\right)^2} \\
&\quad + \frac{\pi}{c_6\lambda nT}(\sum_{b=1}^{\lambda nT}\mathbf{E}[l_b])^2 - \lambda T
\end{aligned}$$

As in the proof of Proposition 2.5, the case with $\sum_{b=1}^{\lambda nT}\mathbf{E}[l_b] \geq \lambda T/n$ will again prevail. Hence,

$$\begin{aligned}
\frac{4c_3}{\Delta^2}WT\log n &\geq \frac{1}{c_1 n\log n}\frac{(\sum_{b=1}^{\lambda nT}1)^3}{\bar{D}\left(2\sum_{b=1}^{\lambda nT}\mathbf{E}[l_b]\right)^2} \\
&\quad + \frac{\pi}{c_6\lambda nT}(\sum_{b=1}^{\lambda nT}\mathbf{E}[l_b])^2 - \lambda T \\
&\geq 2\left[\frac{\pi}{4c_1c_6 n\log n\bar{D}}(\lambda nT)^2\right]^{1/2} - \lambda T \\
&= 2\sqrt{\frac{\pi}{4c_1c_6}\frac{\lambda^2 nT^2}{\bar{D}\log n}} - \lambda T.
\end{aligned}$$

When $\bar{D} = o(n)$, the first term dominates for large n. Hence,

$$\begin{aligned}
\frac{4c_3}{\Delta^2}WT\log n &\geq \sqrt{\frac{\pi}{4c_1c_6}\frac{\lambda^2 nT^2}{\bar{D}\log n}} \\
\lambda^2 &\leq \frac{\bar{D}\log^3 n}{n}\frac{64c_1c_3^2c_6W^2}{\pi\Delta^4}.
\end{aligned}$$

Q.E.D.

Notes

1. We use the following notation throughout:

$$\begin{aligned} f(n) = o(g(n)) &\quad \leftrightarrow \quad \lim_{n\to\infty} \frac{f(n)}{g(n)} = 0, \\ f(n) = O(g(n)) &\quad \leftrightarrow \quad \limsup_{n\to\infty} \frac{f(n)}{g(n)} < \infty, \\ f(n) = \omega(g(n)) &\quad \leftrightarrow \quad g(n) = o(f(n)), \\ f(n) = \Theta(g(n)) &\quad \leftrightarrow \quad f(n) = O(g(n)) \text{ and } g(n) = O(f(n)). \end{aligned}$$

2. Note that changing the shape of the area from a square to a circle or other topologies will not affect our main results.

3. Here we have excluded *probabilistic* scheduling policies. Otherwise, $\mathcal{F}_t$ should be augmented with a σ-algebra that is independent of node mobility in future time slots.

4. A similar observation is used in [1] except that they take a receiver point of view.

5. An assumption we have used here is the *separation of time scale*, i.e., we assume that radio transmissions can be scheduled at a time scale much faster than that of node mobility. Hence, each message can be divided into many smaller pieces and the transmissions of different pieces can be pipelined to achieve maximum throughput [1]. We also assume that the overhead of dividing a message into many smaller pieces is negligible.

References

[1] P. Gupta and P. R. Kumar. The Capacity of Wireless Networks. *IEEE Transactions on Information Theory*, 46(2):388–404, March 2000.

[2] M. Grossglauser and D. Tse. Mobility Increases the Capacity of Ad Hoc Wireless Networks. *IEEE/ACM Transactions on Networking*, 10(4), August 2002.

[3] M. J. Neely and E. Modiano. Capacity and Delay Tradeoffs for Ad-Hoc Mobile Networks. *submitted to IEEE Transactions on Information Theory, available at http://www-rcf.usc.edu/~mjneely/*, 2003.

[4] S. Toumpis and A. J. Goldsmith. Large Wireless Networks under Fading, Mobility, and Delay Constraints. In *Proceedings of IEEE INFOCOM*, Hong Kong, China, March 2004.

[5] R. Durrett. *Probability : Theory and Examples*. Duxbury Press, Belmont, CA, second edition, 1996.

[6] M. E. Woodward. *Communication and Computer Networks: Modelling with Discrete-Time Queues*. IEEE Computer Society Press, Los Alamitos, CA, 1994.

[7] A. E. Gamal, J. Mammen, B. Prabhakar, and D. Shah. Throughput-Delay Trade-off in Wireless Networks. In *Proceedings of IEEE INFOCOM*, Hong Kong, China, March 2004.

[8] G. Sharma and R. R. Mazumdar. Delay and Capacity Tradeoffs for Wireless Ad Hoc Networks with Random Mobility. *preprint available at* http://www.ece.purdue.edu/~mazum/, October 2003.

[9] N. Bansal and Z. Liu. Capacity, Delay and Mobility in Wireless Ad-Hoc Networks. In *Proceedings of IEEE INFOCOM*, San Francisco, CA, April 2003.

[10] S. Diggavi, M. Grossglauser, and D. Tse. Even One-Dimensional Mobility Increases Ad Hoc Wireless Capacity. In *ISIT 02*, Lausanne, Switzerland, June 2002.

[11] E. Perevalov and R. Blum. Delay Limited Capacity of Ad hoc Networks: Asymptotically Optimal Transmission and Relaying Strategy. In *Proceedings of IEEE INFOCOM*, San Francisco, CA, April 2003.

[12] M. Gastpar and M. Vetterli. On the Capacity of Wireless Networks: The Relay Case. In *Proceedings of IEEE INFOCOM*, New York, June 2002.

[13] P. Gupta and P. R. Kumar. Towards an Information Theory of Large Networks: An Achievable Rate Region. *IEEE Transactions on Information Theory*, 49(8):1877–1894, August 2003.

Chapter 3

SELF-OPTIMIZATION IN SENSOR NETWORKS*

Bhaskar Krishnamachari, Congzhou Zhou and Baharak Shademan
Department of Electrical Engineering-Systems
University of Southern California, Los Angeles, CA, 90089
{bkrishna,congzhoz,shademan}@usc.edu

Abstract Wireless sensor networks are expected to be significantly resource-limited in most scenarios, particularly in terms of energy. In recent years, researchers have advocated and studied cross-layer design techniques as the primary methodology to leverage application-specificity for optimizing system performance. We argue that another powerful design principle is to make sensor networks autonomously learn application-specific information through sensor and network observations during the course of their operation, and use these to self-optimize system performance over time. We discuss several example application scenarios where such self-optimization can be used, including localization, data compression and querying. As an in-depth illustration, we then present details of LEQS (Learning-based Efficient Querying for Sensor networks), a novel distributed self-optimizing query mechanism.

Keywords: Sensor networks, self-optimization, self-configuration, node localization.

1. Introduction

Large networks of embedded sensor devices, each capable of a combination of computing, communication, sensing and even limited actuation, are being envisioned to provide an unprecedented fine-grained interface between the physical and virtual worlds. According to a recent National Research Council report, the use of such networks of embedded systems "could well dwarf previous milestones in the information revolution" [1]. The applications of sensor networks that are being investigated and developed range widely, in-

*This work has has been supported in part by grants from NSF (awards number 0325875, 0347621, and 0435505) and by an education grant from Intel.

cluding scientific environmental monitoring, civil structural health monitoring, industrial process monitoring, and military surveillance.

It is well recognized that particularly at large scale, many of these applications present some hard challenges. In particular, resource constraints pertaining to energy are likely to be dominant when networks of battery-operated nodes are required to operate for years [2, 3]. The hardware as well as software protocols for sensor networks are therefore being designed with energy-efficiency as the primary objective [4–9].

A fact that has been exploited to develop energy-efficient designs is that while there are many possible applications of sensor networks in general, any given network is likely to be employed for a particular, reasonably well-defined, application. This enables the development of cross-layer techniques that are designed to exploit application-specific structure. One example is the notion of data-centric networking. In Directed Diffusion [7], for instance, all communication flows are generated in response to queries and notifications pertaining to events. Data-centric routing mechanisms allow intermediate nodes to examine the application-level content of packets, in order to perform in-network aggregation and processing to reduce communication costs [10]. Another example are medium access protocols that take into account application-specific communication patterns. The D-MAC protocol [11], for instance, is designed to be very energy efficient as well as low-latency, for sensor network applications in which communication takes place predominantly over a one-way data gathering tree.

However, while such application-specific pre-engineered design is clearly important for sensor networks, it does not address other significant challenges. In many of the envisioned applications, the sensor network is meant to be completely unattended and often deployed in uncertain, unpredictable and dynamic environments. In such conditions, there is a substantial limit on the extent to which network operations can be optimized prior to deployment.

We argue therefore that another essential, complementary, design paradigm for sensor networks is that they must be *self-optimizing* after deployment. Once the network becomes operational, the different sensor network protocols and services must continually observe the environment and take into account information obtained through sensor and network measurements to adapt and optimize their behavior so as to be more energy-efficient. In other words, sensor networks should be autonomous enough to improve their own performance over time through learning.

Let us first consider briefly some examples that illustrate this notion of self-optimization, based on prior and ongoing work. In the next section, we shall then present in detail one of these examples, the LEQS (Learning-based Efficient Querying for Sensor networks) mechanism. Together, we hope, these examples illustrate the need and wide applicability of novel self-optimization techniques in sensor networks.

- **Simultaneous Node-Target Localization:** One service that is essential in many sensor networks applications is node localization, i.e. identifying the geographic location of each node. This is a non-trivial problem when GPS is not available at all nodes due to either cost or other application-dependent considerations (such as indoor or sub-foliage deployment). Much recent work has examined the possibility of providing localization when only a few reference or beacon nodes are available (e.g. [12–15]); however, all of these are techniques that perform node localization at deployment time and leave a residual error in node locations. Intuitively, a self-optimizing approach in this context could provide a mechanism whereby the node locations can be further improved by using observations made during operation. This is precisely the approach taken in [16], which presents an algorithm to combine node localization with a mobile target tracking task in a sensor network. In the algorithm, each observation of the target by sensor nodes adds a geometric constraint on the position of sensor nodes and over time leads to successive improvements in their position estimates.

- **Model-based compression:** Most of the work on data gathering and routing in sensor networks (e.g. [6, 7]) assumes that the primary data communication results from sensor source nodes transmitting raw measurements to sink nodes, possibly with some limited form of data-aggregation. However sensor networks often monitor physical phenomena that have predictability and smoothness to their underlying spatio-temporal structure that can be captured by suitable spatio-temporal models. For example, a sensor network monitoring temperature flows or diffuse phenomena such as chemical concentrations may be well modeled by a partial differential equation (PDE). In such a case, particularly at large scale, it can be substantially more energy efficient to use sensor observations and in-network processing to determine coefficients of a suitable model and transmit model parameters to the sink rather than individual measurements. This is an example of a self-optimizing mechanism, as the number of bits that need to be transmitted through the network to accurately represent the physical phenomenon being monitored can be reduced over time as successive observations help improve the accuracy of the parameterized model. We are currently developing a self-optimizing mechanism that uses distributed Kalman filters to learn PDE model coefficients for different situations [17]. A similar technique has been developed independently and described in another recent paper [18], where the authors present a general kernel-based linear regression mechanism in which also model parameters are computed and transmitted instead of raw measurements.

- **Target Querying:** The end user in a sensor network may be interested in some specific information instead of all data collected within the network. The basic problem of how to extract that information from the wireless sensor networks is then that of querying. Prior work has focused on simple queries can be (i) flooded [7], (ii) forwarded randomly [19, 20], or (iii) routed directed to the object if the path is pre-configured using, for instance, a hash table [21]. However, with queries such as those intended to identify the location of targets that have some predictability in their movements, there is scope for a self-optimizing mechanism. Such a mechanism for querying can use reinforcements to learn over time how queries should be routed efficiently. The LEQS mechanism[1] that we describe in the following is an example of such a technique.

2. The LEQS Mechanism

We now present the LEQS self-optimizing mechanism for target querying. As we shall show, LEQS is suitable in a context where there are repeated queries for objects with somewhat predictable movement patterns. The key insight we exploit is this: if there is an underlying distribution that describes the location of the object, and there are repeated queries for it, it should be possible to "learn" how to query for this object efficiently over time. In this algorithm, sensor nodes maintain weights indicating the probability with which a given query is forwarded to each neighbor. The query response is used to update these weights on the reverse-path, effectively training the network to locate the object efficiently. Before we describe the details of the LEQS mechanism, we state our assumptions explicitly.

2.1 Assumptions

LEQS requires that there be a bidirectional abstraction of the communication links (as may be provided using blacklisting or other techniques). No node localization or geographical information is required. There may be multiple sinks querying for the same objects. The named object (there could be many such objects) being queried for can be found (or a response to the query can be obtained) at only one location in the network at any given time[2], but it may potentially be at one of many locations. There is assumed to be an underlying stationary spatial location distribution for each identifiable, queried object (but this distribution need not be known to the network or to any entity). This assumption about an underlying regularity/pattern in the object's location is crucial to LEQS – there must be an underlying location pattern that can be learned. In the results section, we will investigate both a scenario where the object is

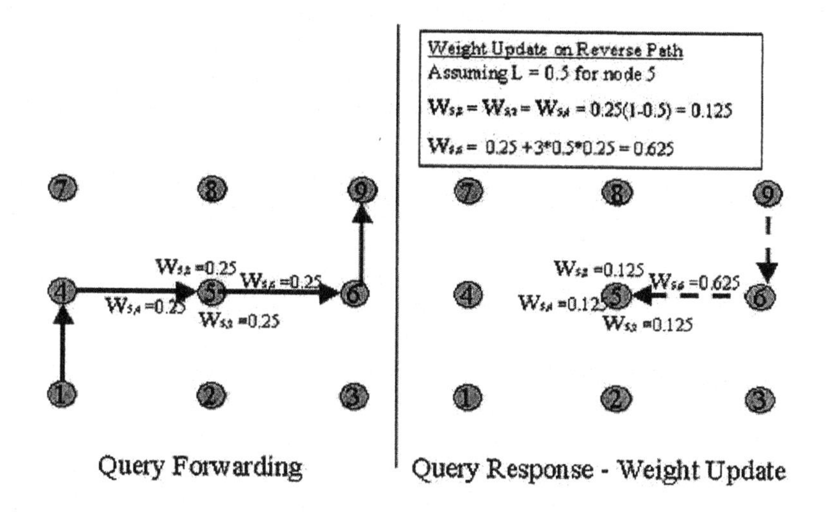

Figure 3.1. Illustration of weight update in LEQS

located in a single location with probability one (which is the most simple case), and the general case when it can be located at one of multiple locations in the network with different probabilities. The LEQS algorithm does not require the use of global unique identifiers for nodes; it only requires unique identifiers for each identifiable object being queried for – a more scalable requirement. Each node communicates through query forward and query response packets only with its immediate neighbors. Because the algorithm is independent of the location where the query is issued, it can be readily used for scenarios in which multiple sinks issue queries for the same object. Finally, it is assumed that multiple queries are issued for the same object over time. Otherwise there will be no opportunity for improving the energy efficiency of the querying via self-optimization.

2.2 Algorithm Description

We now describe the LEQS algorithm. Upon node deployment and setup, each sensor node i identifies its immediate neighbors and sets up a vector of weights $\overline{W_i}$ (one for each identifiable queried object A) in a querying table. Weight $W_{i,j}$ represents the probability that a query for object A that arrives at node i will be forwarded to node j.

Initially, if a given node has k neighbors, each neighbor of the node is assigned an equal weight of $\frac{1}{k}$. At any given time, each query will start from the sink, and with the probabilities denoted by the weights at each node, will be forwarded randomly from node to node. Each node will forward the query to one of its neighbors, except the node it has received the query from, according to their weight. Each node that receives the query checks to see if the queried object is located at that node. A backtracking technique is incorporated to prevent looping[3]. Eventually the query will find its way to the node where the object is located. The response of the query is then sent back directly to the sink on what is essentially the reverse path of this query (using information recorded at each node along the way, bypassing any backtracked branches) and this is when the weight updates occur. Each node i on the reverse path increases the weight of its next-hop h_i (i.e. the next node on the forward path between i and the node where the query terminated successfully). The query response on the return path contains a counter d that is incremented hop by hop, so that all nodes on the reverse path get an indication of how many hops they were from the query termination location. This information is used in the weight update rule, which is described below.

Each node on the reverse path first calculates a learning factor L_i as follows:

$$L_i = \frac{p}{d_i^\alpha}$$

In the above equation, $p \in [0, 1]$ and α are learning parameters that determine the rate of learning as well as the dependence on d_i, the distance (in hops) of node i from the query termination point. Let h_i be the node from which i receives the query response (i.e. the node to which i had originally forwarded the query). Let $N(i)$ be the set of all neighbors of i. Then, the weights at i are updated as follows:

$$\begin{aligned} W_{i,j}(t+1) &= W_{i,j}(t) \cdot (1 - L) \quad , \forall j \in N(i) \backslash h_i \\ W_{i,h_i}(t+1) &= W_{i,h_i}(t) + L \cdot \sum_{j \in N(i) \backslash h_i} W_{i,j}(t) \end{aligned} \tag{3.1}$$

Note that this update rule ensures that the weight of all the neighboring nodes always sums up to 1. We refer to this weight update as the learning process. In case of an object that may be potentially located at one of multiple locations, each with different probabilities, over time, the query will learn to optimize its path to visit all of these locations in turn so that it minimizes the expected number of hops in the query. Also note that the query table at each node includes a row for each identifiable object with a column for each neighbor. Thus the storage requirement per node is $O(kT)$, if k is the (max) number of neighbors per node and T the number of objects being queried.

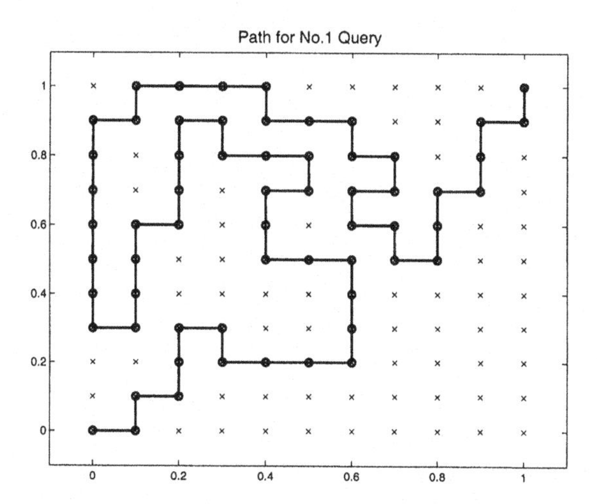

Figure 3.2. Sample run showing how learning improves the query efficiency (single location case) over multiple queries: Query 1.

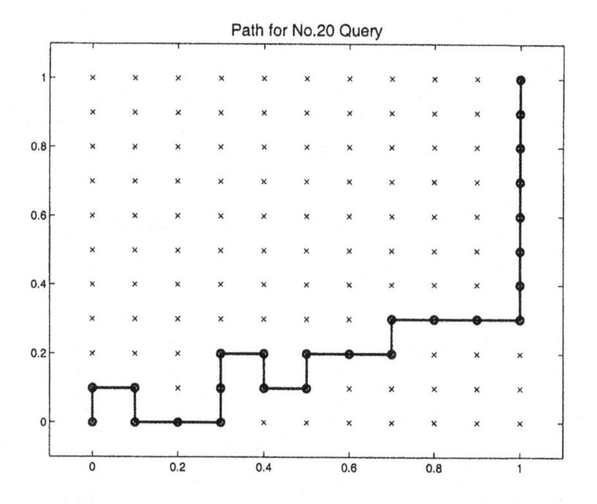

Figure 3.3. Sample run from Figure 3.2: Query 20.

3. Simulation Experiments and Results

3.1 Metrics

We first briefly discuss different metrics that can be used to measure the performance of LEQS. Metrics A and B are shown in simulation results presented in this paper.

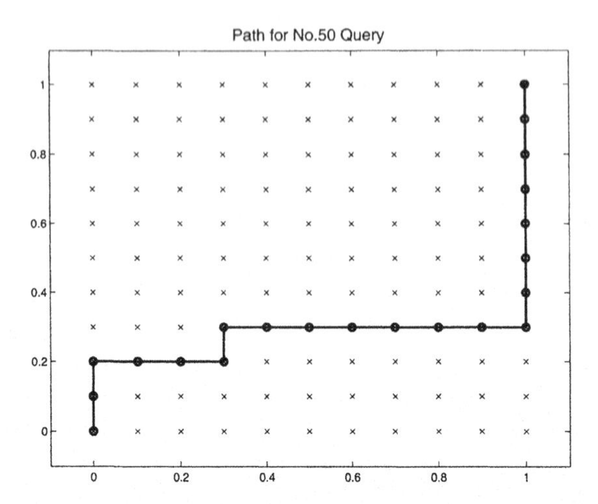

Figure 3.4. Sample run from Figure 3.2: Query 50.

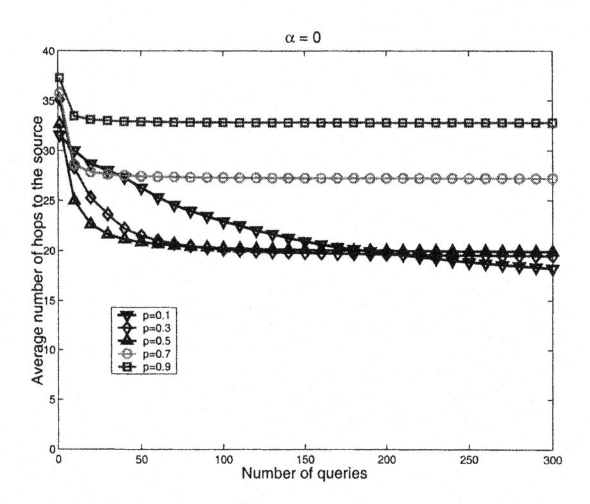

Figure 3.5. Decrease in average cost of querying over time for different learning rates for $\alpha = 0$ for random deployment

A. Average Total Number of Hops (in path from object to sink): This is the expected number of hops that a query response takes from the located object to the sink. (i.e. the number of hops on the reverses path). This can be measured instantaneously for each query, or cumulatively averaged over all preceding queries (which will amortize the higher cost of initially inefficient paths over future, more efficient queries). Note that this one-way metric does not take into

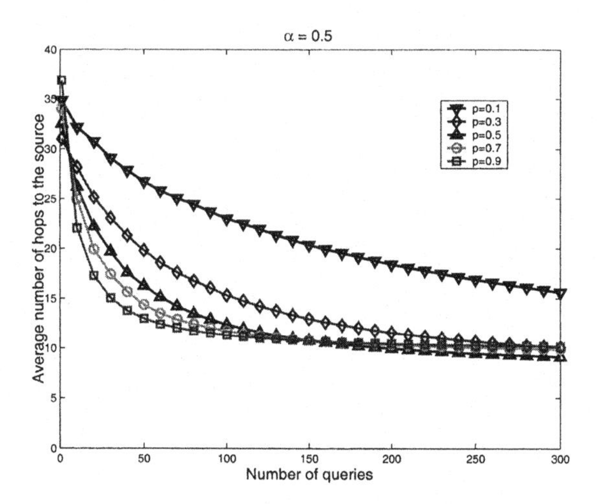

Figure 3.6. Case from Figure 3.5 but with $\alpha = 0.5$

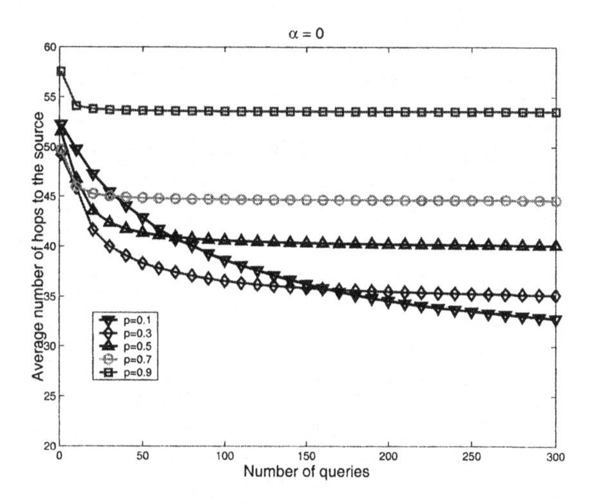

Figure 3.7. Case from Figure 3.5 but with grid deployment

account additional transmissions due to branching/backtracks during the query forwarding phase. Those are best captured by the following metric.

B. Average Number of Transmissions: A related metric is the average number of transmissions required to forward the query. Since query responses are sent in reverse of the original query, the two are almost the same. The difference is that there can be additional overhead of up to 50% in the forward direction due to the random walk branches in initial queries that result in backtracks (to

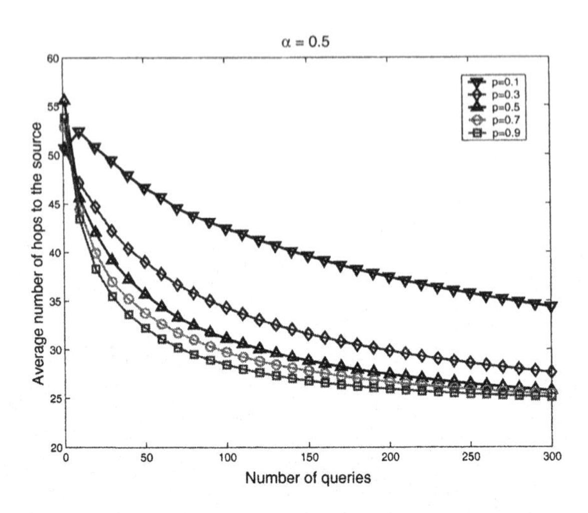

Figure 3.8. Case from Figure 3.7 but with $\alpha = 0.5$

prevent query looping). But the number of such backtracks decreases over time so that eventually the two metrics A and B become identical.

C. Convergence Time: This is the time needed for the network to converge to a weight distribution that does not change (which happens when queries settle to a single possible path). In our simulations, we measure time by the number of queries issued by the sink for the object. This metric is useful in theory, but not explicitly presented in our simulations.

3.2 Simulation Results

We uniformly place 121 static nodes in a 1 x 1 square area. All nodes in the network have the same communication range R. The sink is located at the left lower corner[4]. All figures shown for the cumulative average number of hops are averaged over 20 simulation runs.

Figures 3.2-3.4 depict a single LEQS run for the first, 20th, and 50th queries in case of a single object location. Starting from using random walk (the first query), the network learns through time about the distribution of object locations. The 20th query clearly performs better than the first query. The 50th query traverses the optimal path.

Our initial approach for studying the learning process was to change weights without considering the distance to the object, in other words, setting $\alpha = 0$. Figures 3.5 and 3.7 show the results for this approach. As we can see from the figure, if we change weights fast (for example, p=0.9), we will have a solution fast but that may not be optimal. On the other hand, if we change weights slowly (for example, p=0.1), we will have a solution after a long time but the solution is likely to be near-optimal.

Our second approach is to incorporate the distance from the point where the query was terminated into consideration when changing weights. Figures 3.6 and 3.8 show the results for this approach. As we can see from the figure, if we change weights fast, we will have a solution fast and the solution is reasonably good. Comparing these two approaches, our second approach has a significant improvement in terms of the solution quality except when the learning parameter p is too small. We will further examine the impact of α in detail below.

Figures 3.5-3.8 also confirm that the trends are similar regardless of whether the deployment is random or on a grid. Due to the similarity in their results, we restrict our remaining results to scenarios involving grid deployment.

Figure 3.9 shows that the performance curves for different settings of p can cross-over. If p is chosen to be very high (0.95), LEQS performs well initially as the learning rate is higher; however it may settle down to a non-optimal value, as seen by the fact that the lower learning rate curve eventually provides a lower average cost. This suggests that there is a tradeoff between the speed of convergence and the optimality of the converged solution that can be achieved by a careful selection of this parameter p.

We now consider what happens if the object is not always located at one node, but rather in one of several locations with an underlying probability distribution.

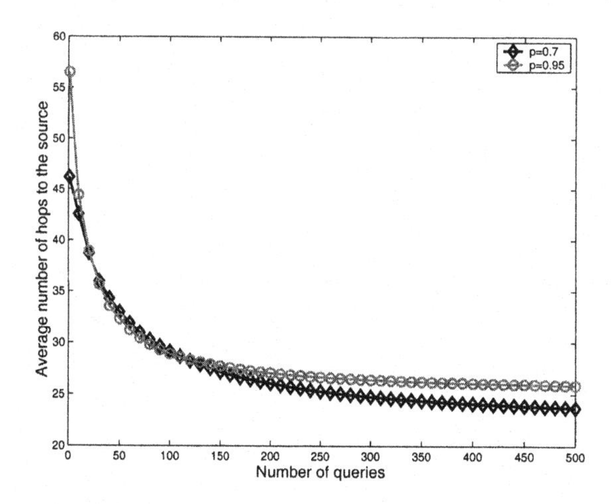

Figure 3.9. Performance of Query Learning for different learning rates; the crossover shows that for lower values of p the learning may be slower but converge to a better value than with high p.

As illustration of the performance of LEQS for multiple location scenarios, we examine some specific examples. Figures 3.10-3.11 show the results for a five-object-locations case. We can see the distribution of the five object

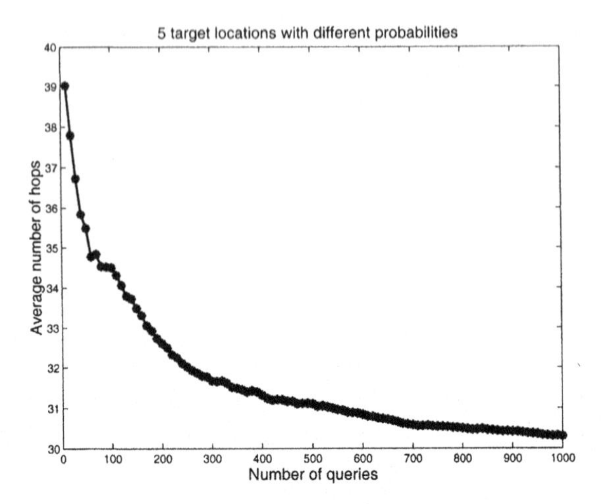

Figure 3.10. Performance of Query Learning.

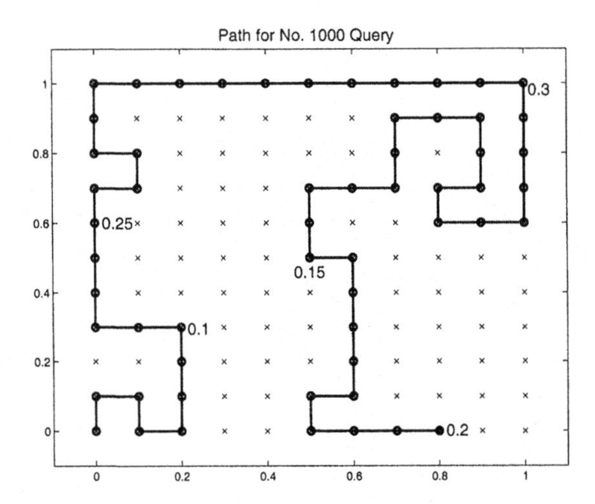

Figure 3.11. A sample run after 1000 queries for scenario involving five possible locations with different probabilities.

locations with the corresponding probabilities. The figure also depicts a sample run for the 1000th query, when the object is located at (0.8,0).

Figures 3.12-3.13 compare the performance of LEQS with respect to random walk, flooding, and the global optimum solution for both single and 5-location scenarios. The performance of LEQS is shown for both the cumulative average (which takes into account the cost of the initial inefficient queries) as well as the instantaneous value of the number of transmissions taken to reach the object.

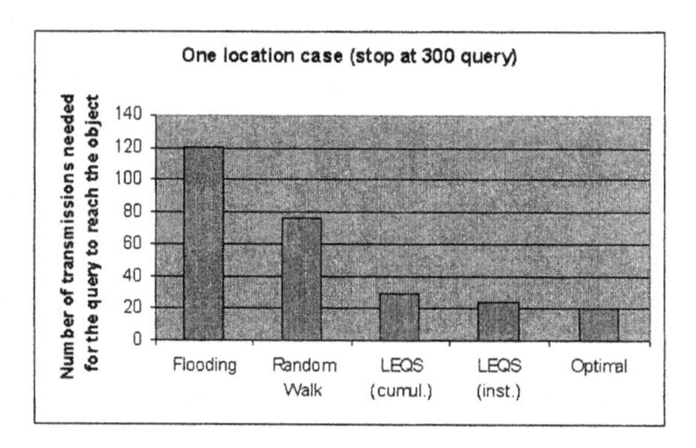

Figure 3.12. Comparison of LEQS, both cumulative (which amortizes initial cost of learning) and instantaneous, with respect to Random Walk, Flooding and the global Optimal solution, for one-location scenario.

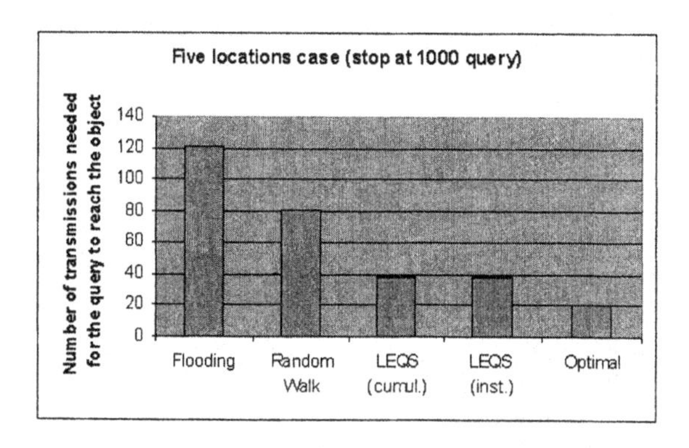

Figure 3.13. Comparison as in Figure 3.12 but for five-location scenario.

The global optimum shown is the optimal sequential solution — this is obtained by performing an exhaustive search that evaluates paths involving visiting each location (all permutations).

In the case of a single object location, the performance of LEQS is near-optimal, and shows that it offers nearly 75% gains with respect to a random walk. In the five-location case, it takes longer to converge to the optimal solution (this has not yet occurred after 1000 queries), but the performance of LEQS still shows more than 50% improvements. Note that these gains can be even higher for larger networks.

In cases where there is a node failure, our scheme has the ability to be self-healing. As long as there exists a path between the sink and the object, the network will find it and adapt to it. Figures 3.14-3.15 show the no-failure cases as well as a scenario in which there is a node failure (node 26) at query number

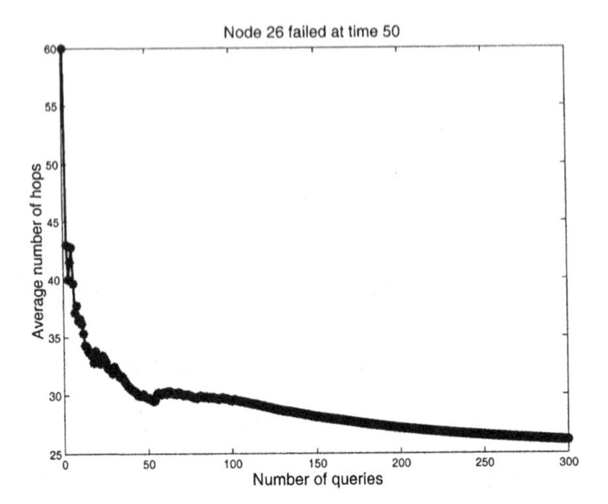

Figure 3.14. Average query cost with respect to query number with a node failure on the query path at query number 50, demonstrating the self-healing nature of LEQS.

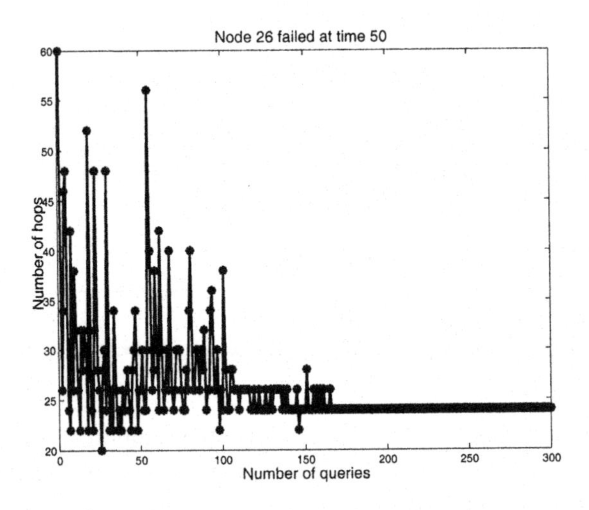

Figure 3.15. Instantaneous query cost with respect to query number with a node failure on the query path at query number 50, demonstrating the self-healing nature of LEQS.

50. Although there is a jump at query number 50 in the figure (seen in both the instantaneous and the average curves) the learning process continues after the failure, and the number of hops to success still keeps decreasing towards the optimal value as the number of queries increase.

Note that although we have not explicitly compared LEQS with flooding or random walks here, these comparisons are implicit – flooding in this network has a cost of $n = 121$ transmissions. A random walk may take about $n/2 \approx 60$

hops to locate the object, and this cost is shown implicitly as the cost of LEQS at query number 1, when no learning has taken place. In the scenarios we considered, we found that LEQS can result in energy efficiency improvements of up to 75% given sufficient learning time and optimal parameter settings.

4. Conclusions

LEQS is an illustrative example of what we argue is a necessary paradigm shift in sensor networks: from pre-deployment engineering and optimization alone to post-deployment self-optimization. This new design paradigm requires that applications and services for sensor networks take explicitly into account sensor and network observations, learning from them in order to continually improve network performance during their operation.

Notes

1. Pronounced "lex", the acronym LEQS comes from the phrase *Learning-based Efficient Querying for Sensor networks*. A more detailed description and analysis of the LEQS mechanism is provided in a technical report [22].

2. This kind of query is useful, for example, to locate identifiable targets in the sensor network region

3. In the LEQS algorithm, every attempt to sending a query to the object results in a success since the walk backtracks if a loop is encountered. With sufficient time, the entire network will be searched in the worst case. A TTL field must be added to the query packet if shorter latency constraints are required, but this may result in query failure. Backtracking does add additional overhead to LEQS, but our simulations confirm that backtracking occurs less frequently as the learning proceeds.

4. Note that while we use a single sink in our experiments, this is not at all a requirement for LEQS — it works equally well with multiple sinks issuing queries for the same identifiable object/target.

References

[1] D. Estrin *et al. Embedded, Everywhere: A Research Agenda for Networked Systems of Embedded Computers*, National Research Council Report, 2001.

[2] D. Estrin, R. Govindan, J. Heidemann, and S. Kumar, "Next Century Challenges: Scalable Coordination in Sensor Networks," *Mobicom*, 1999.

[3] G.J. Pottie, W.J. Kaiser, "Wireless Integrated Network Sensors," *Communications of the ACM*, vol. 43, no. 5, pp. 551-8, May 2000.

[4] G. Asada *et al.*, "Wireless Integrated Network Sensors: Low Power Systems on a Chip," *Proceedings of the 1998 European Solid State Circuits Conference*.

[5] R. Min *et al.*, "An Architecture for a Power-Aware Distributed Microsensor Node," *IEEE Workshop on Signal Processing Systems (SiPS '00)*, October 2000.

[6] W. Heinzelman, A. Chandrakasan, and H. Balakrishnan, "Energy-Efficient Communication Protocol for Wireless Microsensor Networks," *Proc. Hawaii Conference on System Sciences*, Jan. 2000.

[7] C. Intanagonwiwat, R. Govindan and D. Estrin, "Directed Diffusion: A Scalable and Robust Communication Paradigm for Sensor Networks," *ACM/IEEE International Conference on Mobile Computing and Networks (MobiCom 2000)*, August 2000, Boston, Massachusetts.

[8] W. Ye, J. Heidemann, and D. Estrin, "An Energy-Efficient MAC Protocol for Wireless Sensor Networks," *INFOCOM 2002*, New York, NY, USA, June, 2002.

[9] Y. Yu, V. K. Prasanna, "Energy-balanced task allocation for collaborative processing in networked embedded systems," *LCTES*, 2003.

[10] B. Krishnamachari, D. Estrin, and S. Wicker, "Impact of Data Aggregation in Wireless Sensor Networks," *International Workshop on Distributed Event Based Systems, DEBS'02*, July 2002.

[11] G. Lu, B. Krishnamachari and C. Raghavendra, "An Adaptive Energy-Efficient and Low-Latency MAC for Data Gathering in Sensor Networks," *4th International Workshop on Algorithms for Wireless, Mobile, Ad Hoc and Sensor Networks (WMAN 04), held in conjunction with the IEEE IPDPS Conference 18th International Parallel and Distributed Processing Symposium*, Santa Fe, New Mexico, April 2004.

[12] L. Doherty, K. S. J. Pister, and L. El Ghaoui, "Convex position estimation in wireless sensor networks," *Infocom 2001*, Anchorage, AK, 2001.

[13] Andreas Savvides, Chih-Chieh Han, Mani B. Srivastava, "Dynamic fine-grained localization in Ad-Hoc networks of sensors," *Mobicom 2001*.

[14] A.Savvides, H.Park and M. Srivastava, "The bits and flops of the N-hop multilateration primitive for node localization problems," *WSNA '02*, 2002.

[15] S. Simic and S. Sastry, "Distributed Localization in Wireless Ad Hoc Networks," *Memo. No. UCB/ERL M02/26*, UC Berkeley, 2002.

[16] A. Galstyan, B. Krishnamachari, K. Lerman, and S. Pattem, "Distributed Online Localization in Sensor Networks Using a Moving Target," *Symposium on Information Processing in Sensor Networks (IPSN)*, April 2004.

[17] L. Rossi, B. Krishnamachari, and C. C. Kuo, "Hybrid Data and Decision Fusion Techniques for Model-Based Data Gathering in Wireless Sensor Networks," *in preparation*.

[18] C. Guestrin, P. Bodik, R. Thibaux, M. Paskin, and S. Madden, "Distributed Regression: An Efficient Framework for Modeling Sensor Network Data," *Symposium on Information Processing in Sensor Networks (IPSN)*, April 2004.

[19] D. Braginsky and D. Estrin, "Rumor Routing Algorithm for Sensor Networks," *First ACM Workshop on Sensor Networks and Applications (WSNA)*, September 2002.

[20] N. Sadagopan, B. Krishnamachari, and A. Helmy, "The ACQUIRE Mechanism for Efficient Querying in Sensor Networks," *First IEEE International Workshop on Sensor Network Protocols and Applications (SNPA), in conjunction with IEEE ICC 2003*, Anchorage, AK, May 2003.

[21] S. Ratnasamy, B. Karp, L. Yin, F. Yu, D. Estrin, R. Govindan, and S. Shenker, "GHT – A Geographic Hash-Table for Data-Centric Storage," *First ACM International Workshop on Wireless Sensor Networks and Applications (WSNA)*, September 2002.

[22] Bhaskar Krishnamachari, Congzhou Zhou, Baharak Shademan, "LEQS: Learning-based Efficient Querying for Sensor Networks," USC Computer Science Technical Report CS 03-795, 2003.

Chapter 4

FROM INTERNETS TO BIONETS: BIOLOGICAL KINETIC SERVICE ORIENTED NETWORKS

The Case Study of Bionetic Sensor Networks

Imrich Chlamtac, Iacopo Carreras and Hagen Woesner
CREATE-NET Research Consortium
Trento, Italy
{chlamtac,iacopo.icarreras,hwoesner}@create-net.it

Abstract As the trend toward ubiquitous and pervasive computing continues to gain momentum, new networking paradigms need to be developed to keep pace with the needs of this emerging environment. In the near future we can expect the number of nodes to grow by multiple orders of magnitude as tags, sensors, body networks etc., get fully integrated into the communication superstructure. Not only will the amount of information in these all-embracing pervasive environments be enormous and to a large degree localized, but also the relaying needs for maintaining an end to end reliable 'always on' networks, as we know today, will be, for the vast majority of the pervasive users beyond their resource capabilities, and in addition redundant, considering the needs of the personalized services dominating these networks. The ambiance within which these nodes will act will be intelligent, mobile, self-cognitive and not limited to machine to machine communication. In these networks the end to end concept of always on communication that formed the basis of the Internet for the last three decades will become passe. The next 'Internet' frontier will be the challenge of adjusting to the omnipresent, intelligent and self-cognitive networking environments which are becoming an integral part of the new societal reality.

These networks will be characterized by their ability to continuously adjust to the environment, by their, often intelligent, users mobility and by their architecture being defined by the services at hand. Compared to classical networking architectures, in these environments communication networks will locally self-organize when the opportunity or need arise, will adjust and evolve over time and will cease to exist when obsolete with respect to a given service or application. Given the need for continuous adjustment and evolution, the user-service focus, the mobility, and the evolving distributed intelligence and user cooperation not imposed by arbitrary service independent protocols, but sought by those users sharing a common interest in service, the underlying networks behavior may seem closer to living organisms. Secondly, the biological model fits well the concept of services that define the network at any given point in time. In this context the biological model can naturally define a service as the organisms chromosomes, which not only store genetic information but participate and are subject to rules of evolution, leading to the ability of the service to evolve and adjust to changing environments. Furthermore, given the complexity and size of user population - unpredictable user needs, to remain competitive and 'survive'. Thirdly, most of the information exchange in these networks occurs locally between users on the move, or kinetic, users. The intelligence, resources and mobility of the kinetic users who carry services in their chromosomes, makes it possible to mimic the interaction occurring in the biological world where mating and mutations lead to interactions that form the basis of evolution. Using this analogy it becomes possible to derive network elements as well as behavioral rules from nature to replace current protocol concepts, and thus avail the communication system of the benefits of natural evolution perfected over millions of years. With this approach the mobile pervasive environments may witness a paradigm shift in the way we view networking is perceived, rather than contemplating a gradual evolution of the classical protocols and their adaptation to the 'next generation network' which cannot deal with the growth and nature of the emerging pervasive environments.

In this chapter we first define the principles and rules of the BIOlogical kiNETic Service centered (BIONETS) or Bionetic networks. We then consider their application in the case of wireless sensor networks and evaluate their effectiveness considering a real life parking service scenario.

Keywords: Genetics, evolution, bio-networking, pervasive computing, mobile ad-hoc networks, wireless sensor networks.

1. Introduction

As the trend toward ubiquitous and pervasive computing gains momentum, the underlying networking concepts need to keep pace with it. The current networking paradigms date from three decades ago. Whether local [7], or long haul [10] levels, whether using wireless [15] or optical [5]technologies, Internet today follows the fundamental concepts of layered communication where the system is set up on a quasi permanent basis to guarantee end-to-end connectivity and any required level of services that might be requested at some undefined future time.

As Internet itself is evolving into being the transport mechanism of 'all that communicates' and all that moves [11], it needs to evolve along with the emerging pervasive, omnipresent environment it will have to support. By pervasive computing we now mean an invisible halo of computing and information services that persist regardless of location, including, in addition to the standard devices that form most Internet nodes today, also wearable computers, smart home and smart building devices, tags identifying virtually every object we come in contact with and sensors providing information about our environment, as well as local intelligence and evolving self-cognitive presence. To this we need to add mobility, an essential feature of our existence and mode of operation.

As aptly stated in [13] 'Taking a step forward, new communication paradigms should focus on the development of intelligent, self-cognitive networks that no longer act as a means to simply propagate information from one machine to the other, but become a living partner of individual and societal activities'. In this context, it is foreseen that we will move toward the development of cognitive situated networks that will play a significant role in person- and society-focused communications. One key area in this field is the development of cognitive sensor networks that will be able to bridge the physical world with the digital world [13][3], and to promote health [21][18][14], safety [14], productivity and knowledge through communication of the network with the environment. Cognitive sensor networks will be built with the deployment of large numbers of autonomous sensor and actuator nodes. Using a large number of specialized sensors and actuators in a dense network we will be able to acquire localized and situated information of certain metrics gathered from the physical and/or digital environment. These networks will use this collection of situated measurements in order to recognize and control certain events in the physical and/or the digital world, for promoting health, safety, communications and knowledge. The intelligence of such networks does not lay on the nodes themselves, which have very limited recourses and capabilities, but in the size and complexity of the network. In fact, the architecture of the network could be seen as a programming language that is used to solve a problem [17].

With the number of nodes growing by multiple orders of magnitude as tags, sensors, body networks etc., get fully integrated into the communication superstructure, the classical protocols will therefore not be able to provide viable support or be executable on the myriad of tiny inexpensive devices in terms of the multilayered protocol stack to be processed or the relaying burden imposed on these devices by standard networking approaches. In fact the need for these devices to be extremely low cost and expendable, while executing an increasing number of instructions and relying increasing amounts of data, leads to a fundamental paradox in the context of classical Internet operation. Furthermore, not only will the amount of information in the all-embracing pervasive environments be enormous, it will be to a large degree localized. Hence the relaying requirements imposed on the vast majority of the pervasive users and devices in order to maintain an end to end reliable 'always on' network will be not only beyond their capabilities but also, given the localized nature of services and device/network intelligence, will be redundant.

The ability to provide services locally based on the changing interests of users, the continuously varying resources available as well as the diverse communication capabilities of the nodes, means that the new network needs to be able to adjust to ever changing local conditions. Put differently, the ability to evolve, becomes as fundamental to the operation and success of a service in this new environment as the concept of end to end virtual circuits and end to end datagram delivery was in the communication networks of the last three decades. The success of this network will reside in the ability of the users devices to communicate in and with the surrounding environment and to collaborate toward a common goal leading to the successful **self** provisioning of a service which evolves over time, is responsive to changing demands and needs to stay competitive (survive).

Thus, while reliability and end to end connectivity have been the primary focus of the Internet protocols so far, in the new pervasive paradigm it is important to have a network that imposes minimal device (user) communication requirements, is capable to autonomously adjust to the environment, while providing the maximum benefit at the service level. The principal (if not unique) role of this network is to support execution of services for intelligent mobile users and environments. These new networks will locally organize when the opportunity or need arise, will adjust and evolve over time and cease to exist when obsolete with respect to a given service or application, providing only as much support as needed by the application and minimizing effort/cost to expend minimum energy required for the task. In these networks a user can be expected to contribute to the 'common goal' of creating a reliable communication environment only if driven by local interest (or preservation) defined as the successful delivery of its own services.

These networks can therefore be expected to behave more as living organisms than the classical Internet, where users and devices evolve and modify their behavior according to the environment using the rules of genetics and evolution dictated by the task to be executed (service), in the common goal of survival (successful service evolution and delivery), interacting when benefiting the individual with no overall concept of network, unless needed by the individual, and no communication or existence of a 'network concept' can be expected beyond its usefulness, i.e. the utility of a service being executed. We therefore refer to them as *BIOlogical kiNETic Service Oriented networks*, bionetic networks of BIONETS for short.

In summary, given the need for continuous adjustment and evolution, the user-service focus, the mobility, and the evolving distributed intelligence and user cooperation not imposed by arbitrary service independent protocols, but sought by those users sharing a common interest in service, we are motivated to use the genetic evolution model to define the "'elements"' of the network. Secondly, the biological model fits well the concept of services that define the network at any given point in time. In this context the biological model can naturally define a service as the organisms chromosomes, which not only store genetic information but participate and are subject to rules of evolution, leading to the ability of the service to evolve and adjust to changing environments, and furthermore given the complexity and size of user population - unpredictable user needs, to remain competitive and "'survive"'. Therefore we use these rules to set the rules of behavior, or the "'protocols"' of these networks. Thirdly, most of the information exchange in these networks occurs locally between users on the move, or kinetic users. The intelligence, resources and mobility of the kinetic users who carry services in their chromosomes, makes it possible to mimic the interaction occurring in the biological world where mating and mutations lead to interactions that form the basis of evolution. We therefore consider the users as the "'communication channels"' of these networks, the medium for carrying the information and creating the connectivity as dictated by the pervasive environments in which the users operate.

We postulate, that using this analogy, it becomes possible to derive network elements as well as behavioral rules from nature to replace current protocol concepts, and thus obtain the benefits of natural evolution perfected over millions of years in the biological world.

In the next section we define BIONETS and propose a set of communication rules governing them indicated by evolution and genetics. On the principle that no theory is valid unless proven in practice, we demonstrate, in the subsequent section, how the concept of BioNETS can be applied to solve the problem of cost effective wireless sensor networking. Wireless Sensor Networks (WSN) have been investigated in their various forms for over a dozen years. Ranging in applications from the detection, temperature or chemical substances in the field to smart buildings, health and environment, security, identification, alarm, control, to detection, and in cooperation with an intelligent coordinator their use is almost unlimited. Nevertheless, except for military applications WSNs have been around mostly only as research proofs of concept. One of the main reasons for this situation is the cost barrier and the contradiction between the need to produce extremely low cost sensors, while requiring them to execute complex networking functions in the context of classical Internet communication models. Creating a sensor network in this context requires that the sensors themselves provide the communication and connectivity needed for a reliable transport mechanism to support 'whatever application may be temporarily needed'. Given a sensor node range, the number of sensors needed to build a connected sensor network is prohibitive, leading to a market entry barrier that is too high for most customer application. Moreover, the processing, communication power and energy required of the sensors nodes to maintain a standard protocol suite, exceeds their specification and contradicts their price expectations.

We show that by applying the BIONETS principles to wireless sensor networks it is possible to obtain a network in which sensors only serve to collect information, where the 'backbone' is provided by the intelligent mobile user, where the networks is in the 'wake' state and energy is spent only when execution of a task demands it. By eliminating in this way the need for sensor connectivity for end to end communication, for sensor network organization, etc., it becomes possible to dramatically reduce the cost of individual sensors and of the sensor network, thus removing the prohibitive cost barrier of sensor networks to customer world. We term our demonstration of those principles in the context of a sensor network *Bionetic Sensor Network*. We further observe and show that this new BIONETS networking paradigm in the context of a WSN can be seen as an extension of a trend toward hierarchical sensor networks that exploit the mobility of nodes to spread information, thus further validating the bionetic approach through current independent trends. Using a simple example of a parking space finding system covering a city, we demonstrate the idea of networking sensors without a sensor network, following the BIONETS principles.

2. BIONETS Definition

The users form nodes of the network and create the network communication mechanism through their movement, collection of information from the environment and the exchange of this information with other users. The users provide the necessary resources in terms of energy, computational capabilities and bandwidth, as well as the intelligence that defines the actions during information exchange and service execution, needed to execute the service. By using these simple mechanisms together with rules governing the exchange and manipulation of information based on genetics and evolution, the information gets stored, spread through the network, and service gets executed without a global (common) knowledge of the information or the network.

We define the BIONETS information processing rules by observing the similarities between the organic world and the role of a communication system in a mobile pervasive computing world and then creating a mapping between the two. We postulated that this approach may yield similar benefits that years of evolution have brought to living organisms allowing them to react, adapt and evolve together with their environment.

2.1 Drawing the parallel between genetic and the BIOlogical Opportunistic NEtworkS

In a genetic scenario, the principal actors are:

- Cells or other basic organisms

- Chromosomes found within a cell contain DNA storing the genetic information of an organism
- Genes that contain a genetic coding that possesses the building blocks information of an organism
- Genes can be classified as dominant or recessive

The mapping between the genetic model and the networking model is obtained by:

- Users are associated with cells or higher level intelligent consumers of information
- Service/environment related information is carried in users' genes
- A gene is *dominant* or *recessive* depending on its value and effect on the service
- Collection of all service related genes forms a chromosome associated with a service the user is participating in
- Network information exchange is associated with cell interactions, or mating, and interactions with the environment is associated with mutation
- Network state transformation is associated with evolution phases occurring through mating and mutations

The BIONETS network rules of behavior and evolution, substituting classical protocols, are derived from the preceding correspondence and obtained by a mapping from basic genetic evolution laws. Specifically, we consider all service relevant information to be stored in and represented by genes. The collection of a specific service related data constitutes a chromosome. The combination of this chromosome stored information together with user actions, dictated by "innate" cell programming or higher level intelligence, constitutes the service execution. The evolution and adaption occurs in the transformation of the information and, as in biology, through either the union processes between two or more distinct chromosomes of two different user, or through genetic changes, mutations occurring due to the effects of the environment on the genetic information stored. The evolution and the behavior of the users over the lifetime of the service are therefore strongly influenced by the environment in which they are moving, the random encounters with other organisms and their own "intelligent" or "coded-behavior" actions taken in response to their own needs in availing themselves of a given service. Furthermore, in this process, each user has the freedom to apply his own values to actions as a function of his needs, and extant conditions.

Looking closer at the rules of interaction, union among cells (users) occurs between any encountering pairs according to the Mendel's laws. Corresponding genes (information) of two cells take part in this union. The newly generated genes (information) within the chromosomes (services) will be the combination of the genes (information) of the two cells (users). At the end of the interaction the dominant genes will regulate the cells (users) behavior while the recessive genes will be part of the cell's DNA, expressing themselves, potentially, in the behavior of the offsprings. As in biology, repeats of the same gene are prone to reduction in number and the most recently used gene is the most likely to be kept. In the same way duplicate information (genes) will be deleted and only the latest service used information will be kept. According to the second Mendel's Law, genes on different chromosomes of the same cell participate independently in the union process. Similarly, information belonging to different services can be exchanged independently in the union process.

The mating process forces the choice between different partners. The partner is chosen based on attractiveness defined as the type of information (value) of most relevance to the user. Once the best partner is chosen, the mating process takes place and new chromosomes are generated from the users participating in the union. The concept of attractiveness strongly depends on the information the users have (genes and genotype). As in genetics the 'attractiveness' is therefore regulated by the rules that enable the organisms to better adapt to the environment and to survive, similarly the 'attractiveness' in the network is regulated by principles that provide a better user satisfaction at the service level.

The environment plays a crucial role in the cells transformation. Organisms evolve from one stage to another during their life, as their genes interact with the environment at each moment of their life history. The interaction of genes and environment determines the nature of the organism and the way it behaves.

To systematize changes between genes (which are inherited) and developmental outcomes (which are not), genetics makes the fundamental distinction between the *genotype* and the *phenotype* of an organism. Similarly in the network it is possible to distinguish between the information (genes and therefore genotype) that users (organisms) carry and the users behavior (phenotype), which depends on the environment the users encounter in their life and from their internal information (genotype). Genetics defines a *norm of reaction of a genotype*, basically a table showing different organism behaviors in different environments. Similarly in a BIONET we can define different user behaviors (phenotypes) in the different environments in which it moves.

To consider the utility of the BIONETS concept, in the remainder of this chapter we apply the bionetic approach to a Wireless Sensor Network (WSN). We study how through a bionetic WSN we can provide the defined service provided by sensor network while eliminating sensors and sensor network complexity as well as the need for sensor density beyond that required by the service. We term this type of WSN the Bionetic Wireless Sensor Network (BWSN). We then evaluate the expected performance of a BWSN network for a given parking application. In order to make the rest of the chapter self contained we first provide a brief definition of existing WSN architectures.

3. Wireless Sensor Networks

The common general concept of WSN today is that of sensors not just as the source of data of any kind, but also as nodes participating in the transport of data, typically in a multi-hop fashion from the point of origin to the point of use, in most cases, a central server that processes the collected data. To perform this function, all sensors are addressable from the network, communicate with each other, and form a communication network.

3.1 Flat Wireless Sensor Networks

A flat wireless sensor network architecture is an homogeneous network, where all nodes are identical in terms of battery and hardware complexity, except for the sink which acts as a gateway being the one responsible for forwarding the collected information to the final user. Depending on the service and the type of sensors, the sensor density need to be very high (several sensor nodes/m^2). According to [4] for any loss index n, the energy costs of transmitting a bit in a multi-hop fashion can always be made linear with distance, while the signal energy decays as the 4th power of distance. These considerations lead to a multi-hop communication for the flat architecture. A flat wireless sensor network, composed of a large number of nodes (up to millions) leads to a number of challenges in terms of network management and organization, including the management of the sensor nodes, the gathering of information, throughput, routing, energy optimization, etc. When dealing with these issues in the presence of a very large number of sensor nodes, scalability becomes critical. Both routing and MAC need to manage and organize a high number of network nodes and in a highly energy efficient manner. Hierarchical or cluster-based routing is generally adopted whenever scalability both at the MAC and at the network layer is needed.

3.2 Hierarchical Sensor Networks

A hierarchical architecture [19]has been proposed to reduce the cost and complexity of "‘most of the sensor nodes" by introducing a set of more expensive and powerful 'sensor' nodes, thus creating an infrastructure that offloads many of the networking functions from the majority of simpler and lower cost sensors. The hierarchical architecture consists of multiple layers: a sensor layer (SN), a forwarding layer (FN), and Access Point layer(AP). Sensor nodes can only sense the environment and communicate the collected information to the forwarding nodes. Forwarding nodes communicate among themselves and with Access Points, the highest tier in the network and the communication points with the backbone.

To further reduce energy consumption, in particular in sparse sensor networks, in [16] a 3-tier architecture has been proposed, in which mobile agents that periodically pass by the sensors, collect their data and finally drop off the collected data to wired access points when in their proximity. Results show that trading off the amount of buffering at both the sensor nodes level and at the MULEs level, acceptable data success rates can be obtained. A similar approach to simplifying the requirement of individual sensor nodes has been proposed in [20]. Termed the Sensor Networks with Mobile Agents (SENMA), ALOHA protocol is used for communication between the sensors and powerful mobile agents eliminating the need for multihop sensor communication and reducing energy consumption.

In summary, in all instances, in building a conventional WSN the resulting node architecture becomes similar to the traditional Internet node[2]. This means that a multi-layer networking stack is implemented in each sensor node, that allows not only for some (limited) processing, but also for the routing of data and eventually even for reliable transport protocols such as TCP. The need for sensor to sensor communication, or network connectivity, results in huge number of devices to cover even a small to medium size city. Whereas sensors in almost all applications, e.g measuring pollution or other environmental, health, risk or security environments,only one sensor per several, potentially hundred, square meters is required. Alternatively communicating with a backbone network such as 802.16 similarly requires a multi level protocol stack, range, battery and complexity. Sensor nodes are battery driven and therefore operate on an extremely frugal energy budget. Energy optimization becomes crucial in maximizing the lifetime of the entire network and has therefore been long considered a key design objective in research related to wireless sensor networks [8]. To cost effectively deploy sensors in a conventional network scenario, sensors have to come at an ultra-low price, their deployment should be easy (like *smart dust*[12]), price per sensor should be below $1, and their battery should be small, efficient and able to support the transmission of millions of control and data messages per minute to support data transmission, self organization, route discovery and maintenance, clustering, routing, TCP, and relaying of information, all growing with the size of the network. Not surprisingly in the real world, sensor prices are two orders of magnitude away, high degree of integration of the sensors remains a challenge, with battery making up about 50% of the volume of sensor nodes. For a commercial application of sensor networks this means that the cost of the full sensor network and the deployment is prohibitive not only due to the high one time cost of placing thousands and millions of sensor in a city environment, but also due to the high replacement cost.

To allow sensors or any other similar information collecting and generating devices on a large scale to become part of our emerging intelligent pervasive environment in which CAPEX and OPEX are key factors to adoption, a fundamentally different solution is therefore required. We next consider the application of the bionetic approach to sensor networks and evaluate the potential of providing the defined service while eliminating the complexity as well as the need for sensor density beyond that required by the service itself, in order to remove the two major roadblocks to the adoption of current WSN architectures in customer environment.

4. Bionetic Wireless Sensor Networks (BWSN)

The organization of BWSN consists of *mobile user nodes* which are the consumers of the service and *nodes providing information,* or sensor nodes which provide the information relevant to the service. The mobile user nodes play the role of 'cells' of the Bionet, while the sensors (or information providing nodes) are the environment in which operate and are influenced by. Each user can participate and store information about one or more services and spreads the information in the network through his physical movement. Each service related information is associated with one or more chromosomes of the user. Evolution occurs as a consequence of user interactions (mating) and mutations (user and sensor interaction). From each interaction, new genes are created and consequently a different behavior of the user or the user's offsprings becomes possible. The evolution through mating as well as mutations are given by the rules of interaction (i.e. protocol) described in section 2.1.

Changes in the environment are translated into changes in the state of the sensors. These changes are assimilated from the users through mutations and are afterwards shared with other users through matings. The information on the environment will in general be more accurate in the direct surrounding of the sensor, where the service needs to be delivered. The speed with which the information will spread will depend on the number of users in the network, their speed and their pattern of movement, the internal memory of each user, and other factors that can be set to reflex the quality of service needed. We further observe that the peer-to-peer information exchange gives the network some of the mobile ad-hoc network (MANET) properties [9]. However, rather than following the conventional MANET source/destination *data transport* paradigm of the Internet the Bionetic Wireless Sensor Network exchanges information locally, has no other 'protocol' than the one defined above and is created and ceases to exist corresponding to the interest in a given service. In one analogy while MANET will use flooding or other routing protocols to deliver the information from its source to its destination, BIONETS will exchange, process and understand the information via fusion and intelligent processing and exchange it only with mates and environment through a genetically dictated rules. Further, unlike in MANET, in the case of mutual communication based on shared service the bionetic scenario provides the necessary incentive to the customers to share information, as the only mode to satisfy their own service needs. The user's motivation to participate in the network information dissemination is therefore derived from the need of the user to *survive*, i.e. avail himself of the desired service and to produce a measurable benefit from the service he is participating in. This benefit can be quantified as the *perceived quality* of the service and is the measure of participation as a function of individual cost and benefit. Lastly, in the bionetic network scenario no end to end concept exists, contrary to MANET where large amount of work has been done for instance to define routing protocols and deal with their scalability motivated by the need for envisioned end to end delivery. Given the focalized nature of the BIONETS, no scalability difficulty is present.

In this scenario, the major task of the sensor node is to sense the environment and to transmit the sensed information to the user nodes in their vicinity. In the bionetic sensor network few, simple and cheap sensors are therefore positioned in the environment covered by a service in locations where service related information needs to be collected or delivered only. The complexity of the communication protocols is reduced and no Internet protocol stack is adopted on the sensor nodes. No addressing is needed, instead location information, which can be imprinted at the installation time suffice. The complexity of the network resides in the user nodes having the resources to communicate, store, process and exchange information, and to define the optimal policy for executing a service. The subsequent scenario demonstrates the detailed translation of these concepts into an operating network.

4.1 An example: a parking lot scenario

As an example of the environmental monitoring applications we describe a parking application as a possible scenario for the Bionetic Sensor Network realization and evaluation. The basic service of the system is assistance in finding the nearest parking. Each parking spot in the city is equipped with a sensor capable of sensing whether the parking is free or not and its geographical position (or ID). Each mobile user (vehicle) subscribing to the service is equipped with a device that communicates both with the sensor nodes and with other mobile users who subscribed to that service. The users, while moving around the city, collect information on the status (free/not free) of the parking locations they encounter. This information is stored in their devices, together with a time stamp of when the information has been collected. Each gene is represented by a tuple describing the *location of the sensor/time stamp/state of the sensor* at which the location status has been updated. When a user passes a parking sensor it updates the respective gene (mutation). When two users come into communication range they *mate* and exchange their knowledge on the parking lot status they are aware of. If multiple users come into range the best, most desirable, mate is chosen as a function of its overall value of parking information relative to the desired location of the user. The information on the state of all parking spaces stored in the genes of the device of the mobile user, represents the chromosome. At any given time a user can query his device (PDA or similar), asking for the nearest free parking lot in a specific area of the city. Using the information in the chromosome of the parking service the user can then process the information and make an independent decision as to the best available parking area available.

4.2 Performance evaluation

It is intuitively clear that while the network structure in terms of cost as well as simplicity is significantly reduced it needs to be ascertained to what extent can the given approach yield a desired level of service. While the information on any available parking area is not available in 'real time' one can expect that as the number of users grows (where users carrying information related to parking do not have to be the parking customers alone, but can include city buses, trams, taxis or any other agent, including pedestrians moving around the city) the information can start to circulate around the city in a rapid way. Furthermore, while no backbone is necessary, optionally, users may be willing to access some kind of GIS service. If so, a backbone connection can be established in the user device. However, in the following analysis we will consider the worst case scenario, where only parking customers are exchanging information and no other means of communication, aside the ones given by the bionetic communication scenario are available.

In order to validate the effectiveness of the proposed solution we simulate the described scenario and use a simple analytical model to partially validate the simulation. The metric used to measure the performance of the system is the probability of having the correct information about the parking location when the user arrives at it. Termed *average perceived accuracy* it assumes a value of 0 when the expected few parking is occupied, and 1 when available as expected. If there is no information about the sensor in the list of the user node, a 0.5 value is assigned. The accuracy is calculated as the average of these values and is chosen as it can be expected to reflect the quality of service and therefore user's satisfaction.

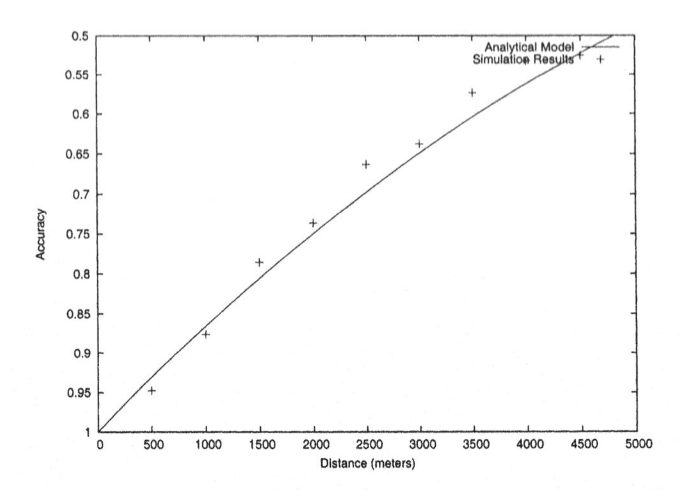

Figure 4.1. Simulation of the simplified analytical model

Simplified Analytical Model. For tractability of analytical modeling consider a single, long bi-directional street with S sensors spaced uniformly along it. The sensors have two states 0 and 1 (this translates as before into a parking slot being occupied or free). The change of the states follows a Poisson distribution with mean λ. Information about the status of sensors is gathered by users driving along the street (at speed v) and by the exchange of information between the nodes.

Consider two nodes that start to drive from opposite ends at a time t_0. They meet in the middle and carry the information of $S/2$ sensors each. We are now interested in the probability of having the correct information about the state of a sensor when arriving at the chosen parking location. The probability for an event X to appear after time t is given by:

$$P(X \geq t) = e^{-\lambda t} \tag{4.1}$$

This means that the state of the sensor did not change within the time t. Assume a certain sensor that has a distance of s from the exchange point. If we want to have a $P = 0.95$ of the correct value, we can calculate the maximum distance in the following, simple way:

$$ln(P(X \geq t)) = -\lambda t \tag{4.2}$$

$$\frac{ln(0.95)}{-\lambda} = \frac{s}{v_1} + \frac{s}{v_2} \tag{4.3}$$

$$s = \frac{v \cdot ln(0.95)}{-2\lambda} \tag{4.4}$$

With a $v_1 = v_2 = v = 13.88\frac{m}{s} = 50\frac{km}{h}$ and a $\lambda = 0.001$ (change in the state of the sensors in average every 17 minutes) this means that a sensor may be 356 meters away from the parking location to know its state with a 0.95 accuracy. The 50% horizon, is on the other hand approximately 4800 meters. Figure 4.1 plots the curve of the analytical model and the simulation points, from the simulation model described next. As shown by this figure the correspondence of the two techniques is high.

Simulation Environment. For studying the performance of the Bionetic Sensor Network in a more complex, two dimensional grid environment we use a detailed simulation model. To this end we developed an evaluation environment using ns-2 [1]. We extended the simplified model described in the previous chapter to a city, where each parking lot of the city center is equipped with a sensor. The change of the state of the parking lot follows a Poisson distribution as in the simplified model, i.e. the time that a parking lot is either free or occupied follows a negative exponential distribution. The communication range of each sensors is taken as 30 m.
Users move around the city and collect information on the state of the sensors (free or occupied) using a CSMA/CA MAC access protocol (specifically the 802.11 ns-2 implementation). When users meet they exchange their knowledge on the state of the parking lot they encountered in their way as described earlier in section 2.1. The mating process has been simulated according to a FIFO (First In First Out) policy so to reflect the relative age of the information. Duplicates genes are deleted afterwards according to their age in the same way as in biology, repeats of the same gene are prone to reduction in number and the most recently used gene is the most likely to be kept (section 2.1). The user buffer is assumed to be large enough to contain the information on the state of all the sensors in the area under consideration. The movement pattern of the users is determined according to a Manhattan Model. Each simulation was run for 10000 seconds (2.7 h) of simulated time. We have run different simulations scaling with the number of users (5, 10, 20, 30, 40, 50), with mean value of the Poisson source (500 and 1000) and with the size of the grid (2000x2000 and 1000x1000). The parameters used in the simulation are summarized in Table 4.1.

Manhattan Mobility Model (MH). The users are moving around in the simulated environment according to the Manhattan Mobility Model. Each mobile node is allowed to move along horizontal and vertical streets.
In our simulation six streets were used, each one of them consisting of two lanes (see Figure 4.2) crossing the grid. At every intersection point of two streets, the mobile user can turn left, right or go straight with probability of 0.25, 0.25 and 50.00. The Manhattan Moving Pattern was determined using the *IMPORTANT* [6] tool.

Simulation Area	1000x1000
Sensors MAC Protocol	802.11
Users MAC Protocol	802.11
Sensors Communication Range	30 m.
Users Communication Range	150 m.
Users Speed	0 - 13.88 m/s
Users Acceleration	1.3 m/s
Simulation Area	1000x1000, 2000x2000
Simulation Time	10000 sec.

Table 4.1. Simulation Parameters

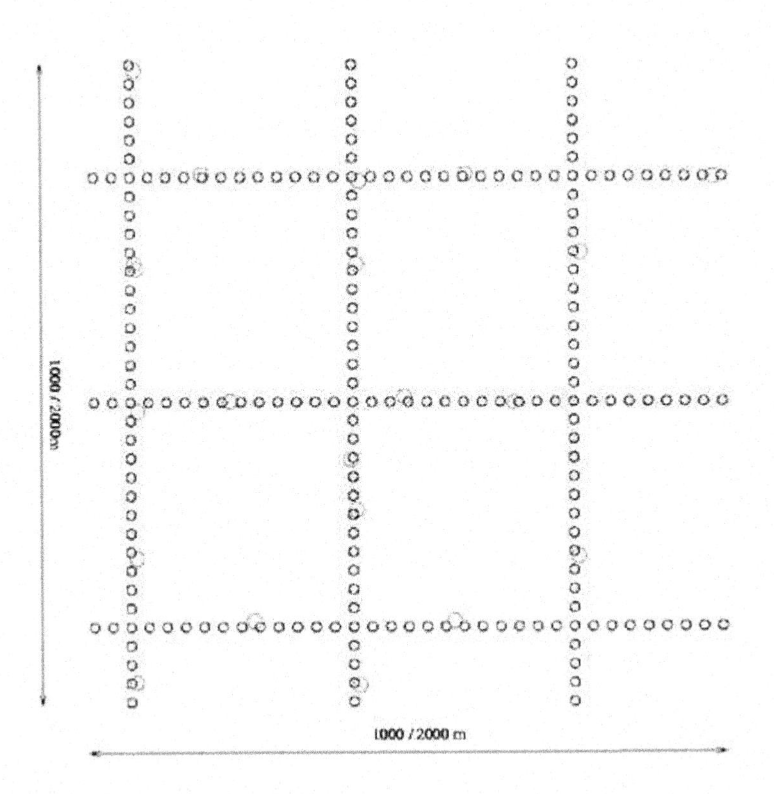

Figure 4.2. Manhattan model of 210 sensor nodes (parking lots).

Number of Users	Average Perceived Accuracy 1000x1000m	Average Perceived Accuracy 2000x2000m
5	0.768802 $\pm$ 0.0383335	0.708716 $\pm$ 0.0355418
10	0.901053 $\pm$ 0.0196254	0.774448 $\pm$ 0.0203497
20	0.941765 $\pm$ 0.0135834	0.864628 $\pm$ 0.0171275
30	0.957831 $\pm$ 0.0110193	0.902028 $\pm$ 0.00854402
40	0.963057 $\pm$ 0.0110836	0.92178 $\pm$ 0.00849648
50	0.973203 $\pm$ 0.014167	0.936325 $\pm$ 0.00893311

Table 4.2. Average perceived accuracy and width of the 95% confidence interval for a 1000x1000m and 2000x2000 grid and a Poisson source with mean of 1000 sec.

Each mobile user is moving with a speed uniformly distributed between 13.88 m/s (= 50 Km/h) and 0 and an acceleration of 1 m/s (3.6 Km/h), corresponding to the typical speed of a car in a city. The next position in each step of the simulation depends on the position of the user its velocity and the position of the other mobile users.
The sensors are uniformly distributed along the streets of the city. In the 2000x2000 grid case the total number of sensors is 210 sensors and in the 1000x1000 grid case the number of sensors is 38.

Simulation Results. Figure 4.3 and Figure 4.4 presents the results of a mean holding time of 1000 seconds (around 17 minutes) with a 1000x1000m and 2000x2000m grid, while in Figure 4.5 and Figure 4.6 we present the results of the same simulation setup for a 500 second holding time.
In each graph we present results with the number of user nodes equal to 5, 10, 20, 30, 40 and 50. As shown in all graphs that after a start up time, the simulation converges to a mean value of around 70% in the worst case and 97% in the best case. The worst case corresponds to 5 users moving in a 2000x2000m grid with a holding time of 500 seconds (Figure 4.6), while the best case corresponds to 50 users moving in 1000x1000m grid with a holding time of 1000 (Figure 4.3). The length of the transient phase and the width of the 95% confidence interval strongly depend on the number of users. The larger the number of users, the faster the information circulates in the grid and the shorter is the start-up phase of the network.

As seen from Figure 4.3, Figure 4.4, Figure 4.5 and Figure 4.6 the accuracy of the information increases with a reduced size of the grid. In a smaller area there is a higher probability that users meet and therefore exchange information. Note, however, that the result is better for 40 nodes on 4 square kilometers than for 10 nodes on one square kilometer, hence further study of behavior is needed.

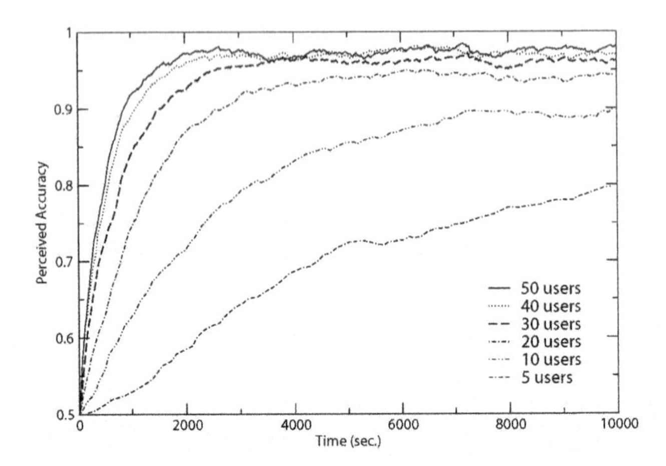

Figure 4.3. Average perceived accuracy with a 1000x1000m area and a mean state holding time of 1000 seconds.

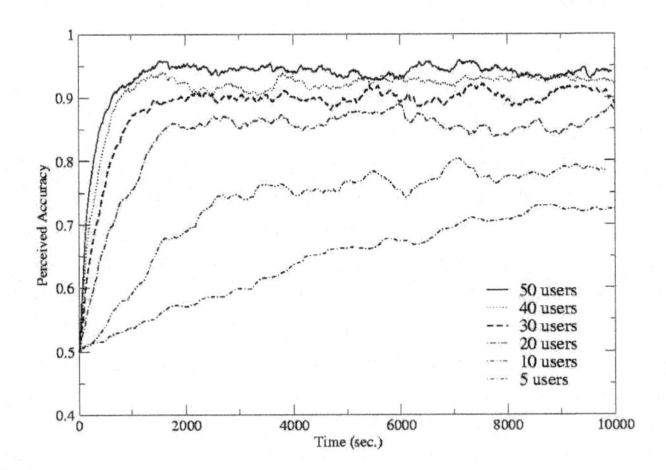

Figure 4.4. Average perceived accuracy with a 2000x2000m area and a mean state holding time of 1000 seconds.

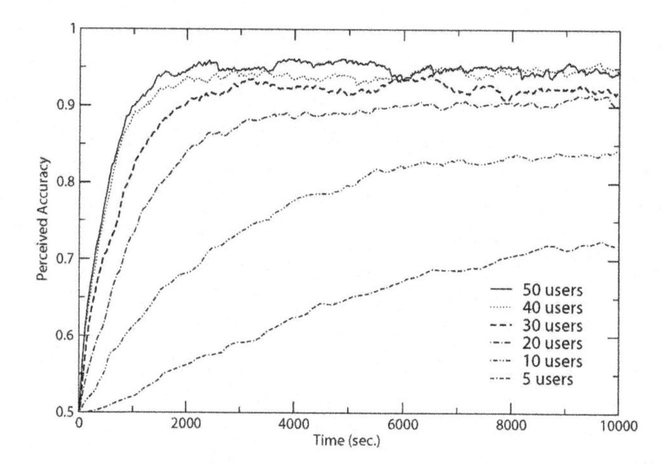

Figure 4.5. Average perceived accuracy with a 1000x1000m area and a mean state holding time of 500 s.

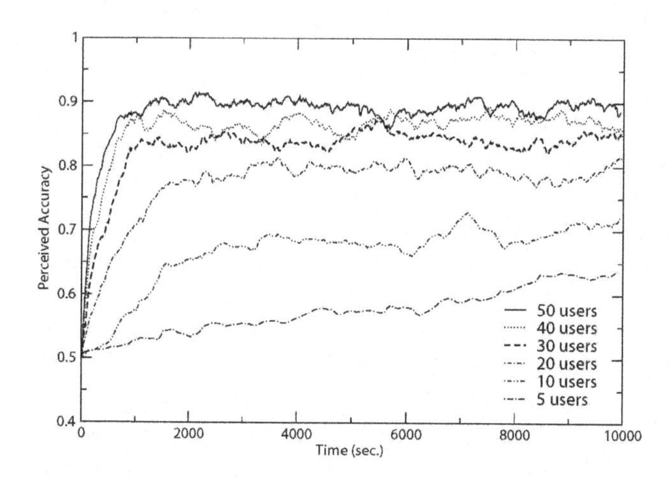

Figure 4.6. Average perceived accuracy with a 2000x2000m area and a mean state holding time of 500 s.

5. Conclusions

The emerging pervasive computing and communications world demands new networking paradigms. In this chapter we proposed BIONETS, a radically new approach to the role and functioning of a communication network needed to support the rapidly growing pervasive environments in which huge number of simple mobile devices populate the supernetwork. We have proposed a model deriving from evolution and genetics that describes the behavior and adaption of a communication system in response to service needs derived from the organic world. We postulated that this approach may yield similar benefits that millions of years of evolution have brought to living organisms allowing them to react, adapt and evolve together with their environment to successfully perform and survive. We have translated this mapping into network behavior rules and showed that by this translation a complete network control can be obtained.

We have applied the BIONETS conceptual model to address a long standing need of developing a sensor network that can function in support of a service at incremental cost, low entry barrier and that fulfills the users need for a given application. In doing so, sensor nodes may be relieved of the burden of multi-hop transport and relaying of data and become essentially passive devices with a very small active (sensing) component. We showed the feasibility of such an approach by considering in detail a simple parking service.

References

[1] The network simulator - ns-2. http://www.isi.edu/nsnam/ns/index.html.

[2] Akyildiz, Ian F., Su, Weilian, Sankarasubramaniam, Yogesh, and Cayirci, Erdal (2002). Wireless sensor networks: a survey. *Computer Networks*, 38:393–422.

[3] Allen, Rebecca (2004). The return of the physical world. "http://www.cordis.lu/ist/fet/comms.htm".

[4] Bhardwaj, M. and Chandrakasan, A. (2002). Bounding the lifetime of sensor networks via optimal role assignments. In *Proc. of IEEE InfoCom, 2002.*

[5] Chlamtac, I. and Jue, J. (2000). *Optical WDM Networks: Future Vision*, volume 9. Kluwer Academic.

[6] F. Bai, N. Sadagopan, A. Helmy (2003). The important framework for analyzing the impact of mobility on performance of routing for ad hoc networks. *AdHoc Networks Journal - Elsevier Science, AdHoc Networks Journal - Elsevier Science*, pages 383–403.

[7] Franta, W. R. and Chlamtac, I. (1981). *Local Networks*. Lexington-Books, Novosibirsk.

[8] I. Chlamtac, C. Petrioli and Redi, J. (1997). An energy-conserving access protocol for wireless communications. In *Proceedings of the 7th annual international conference on Mobile computing and networking.*

[9] I. Chlamtac, M. Conti and Liu, J. (2003). Mobile ad hoc networking: Imperatives and challenges. *Ad-Hoc Networks Journal*, 1(1).

[10] I. Chlamtac, M. El-Zarki (1994). *Introduction to Computer Networks*, volume 9. Marcel Dekker.

[11] I. Chlamtac, Y. Bing Lin and Redi, J. (1998). *When Mobility Meets Computation*, volume 22. Marcel Dekker.

[12] Kahn, J. M., Katz, R. H., and Pister, K. S. J. (1999). Next century challenges: Mobile networking for "smart dust". In *International Conference on Mobile Computing and Networking (MOBICOM)*, pages 271–278.

[13] Kalis, A. (2004). A. kalis,report on fet consultation meeting on communication paradigms for 2020area: Cognitive sensor networks. Technical report.

[14] Keijer, Ulf (2004). Smarthomes. "http://www.cordis.lu/ist/fet/comms.htm".

[15] Lin, Y. Bing and Chlamtac, I. (2000). *Wireless and Mobile Network Architectures*. John Wiley & Sons.

[16] R.C. Shah, S. Roy, S. Jian and Brunette, W. (2003). Data mules: modeling a three-tier architecture for sparse sensor networks. In *Proceedings of the IEEE Workshop on Sensor Network Protocols and Applications (SNPA)*, pages 30–41.

[17] Riguidel, Michel (2004). Communications and computer science in 2020. "http://www.cordis.lu/ist/fet/comms.htm".

[18] Rosenberg, Duska (2004). Understanding mediated communication. "http://www.cordis.lu/ist/fet/comms.htm".

[19] S. Zhao, K. Tepe, I. Seskar and Raychaudhuri, D. (2003). Routing protocols for self-organizing hierarchical ad-hoc wireless networks. In *IEEE Sarnoff 2003 Symposium*, pages 65–68.

[20] Tong, Lang, Zhao, Qing, and Adireddy, Srihari (2003). Sensor networks with mobile agents. In *MILCOM 2003 - IEEE Military Communications Conference*, pages 688–693.

[21] von der Gruen, Thomas (2004). Wireless body area networks in healthcare applications. "http://www.cordis.lu/ist/fet/comms.htm".

Chapter 5

COOPERATIVE ROUTING IN WIRELESS NETWORKS

Amir E. Khandani, Eytan Modiano, Jinane Abounadi and Lizhong Zheng
Laboratory for Information and Decision Systems
Massachusetts Institute of Technology, Cambridge, MA 02139
{khandani,modiano,jinane,lizhong}@mit.edu

Abstract The joint problem of transmission-side diversity and routing in wireless networks is studied. It is assumed that each node in the network is equipped with a single omni-directional antenna and multiple nodes are allowed to coordinate their transmissions to achieve transmission-side diversity. The problem of finding the minimum energy route under this setting is formulated. Analytical asymptotic results are obtained for lower bounds on the resulting energy savings for both a regular line network topology and a grid network topology. For a regular line topology, it is possible to achieve energy savings of 39%. For a grid topology, it is possible to achieve energy savings of 56%. For arbitrary networks, we develop heuristics with polynomial complexity which result in average energy savings of 30% – 50% based on simulations.

Keywords: Wireless, cooperation, routing, energy efficiency, diversity, ad-hoc networks.

1. Introduction

In this chapter, we study the problem of routing, cooperation and energy efficiency in wireless ad-hoc networks. In an ad-hoc network, nodes often spend most of their energy on communication [1]. In most applications, such as sensor networks, nodes are usually small and have limited energy supplies. In many cases, the energy supplies are non-replenishable and energy conservation is a determining factor in extending the life time of these networks. For this reason, the problem of energy efficiency and energy efficient communication in ad-hoc networks has received a lot of attention in the past several years. This problem, however, can be approached from two different angles: energy-efficient route selection algorithms at the network layer or efficient communication schemes at the physical layer. While each of these two areas has received a lot of attention separately, not much work has been done in jointly addressing these two problems. Our analysis in this chapter tackles this less studied area.

Motivated by results from propagation of electromagnetic signals in space, the amount of energy required to establish a link between two nodes is usually assumed to be proportional to the distance between the communicating nodes raised to a constant power. This fixed exponent, referred to as the path-loss exponent, is usually assumed to be between 2 to 4. Due to this relationship between the distance between nodes and the required power, it is usually beneficial, in terms of energy savings, to relay the information through multi-hop route in an ad-hoc network. Multi-hop routing extends the coverage by allowing a node to establish a multi-hop route to communicate with nodes that would have otherwise been outside of its transmission range. Finding the minimum energy route between two nodes is equivalent to finding the shortest path in a graph in which the cost associated with a link between two nodes is proportional to the distance between those nodes raised to the path-loss exponent. Figure 5.1 shows an example of a multi-hop route between two nodes.

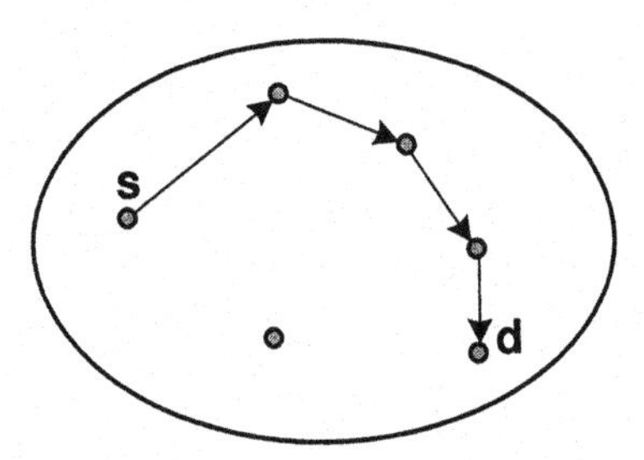

Figure 5.1. Multi-hop Relaying

The problem becomes more interesting once some special properties of the wireless medium are taken into account. In particular, there are three properties of the wireless physical layer that have motivated our work: the wireless broadcast property, the benefits of transmission side diversity, and multi-path fading.

A wireless medium is a broadcast medium in which signal transmitted by a node is received by all nodes within the transmission radius. For example, in figure 5.2, the signal transmitted by s is received by both nodes 1 and 2. This property, usually referred to as the *Wireless Broadcast Advantage* (WBA), was first studied in a network context in [3]. Clearly, this property of the wireless physical medium significantly changes many network layer route selection algorithm. The problem of finding the minimum energy multi-cast and broadcast tree in a wireless network is studied in [3] and [4]. This problem is shown to be NP-Complete in [5] and [6]. WBA also adds substantial complexity to route selection algorithms even in non-broadcast scenarios. For example, this model is used in [8] in the context of selecting the minimum energy link and node disjoint paths in a wireless network.

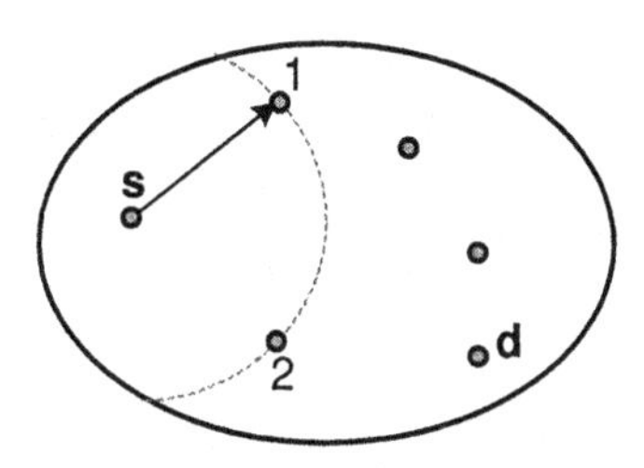

Figure 5.2. Wireless Broadcast Advantage

Another interesting property of the wireless medium is the benefit of space diversity at the physical layer. This type of diversity is achieved by employing multiple antennas on the transmitter or the receiver side. It is well known that transmission side diversity, i.e. using multiple antennas on the transmitter, results in significant energy savings (see [2]). In the network setting studied in this chapter, we assume that each node is only equipped with a single antenna. Hence, a straight forward extension of multiple-antenna results to a network setting is not possible. However, it might be possible that several nodes can cooperate with each other in transmitting the information to other nodes, and through this cooperation effectively achieve similar energy savings as a multiple antenna system. We call the energy savings due to cooperative transmission by several nodes the *Wireless Broadcast Advantage*. An overview of different transmission side diversity techniques is given in [2]. An architecture for achieving the required level of coordination among the cooperating nodes is discussed in [9].

In the problem studied in this chapter, we intend to take advantage of the wireless broadcast property and the transmission side diversity created through cooperation to reduce the end-to-end energy consumption in routing the information between two nodes. To make it clear, let's look at a simple example. For the network shown in figure 5.1, assume the minimum energy route from s to d is determined to be as shown. As discussed previously, the information transmitted by node s is received by nodes 1 and 2. After the first transmission, nodes s, 1 and 2 have the information and can cooperate in getting the information to d. For instance, these 3 nodes can cooperate with each other in transmitting the information to node 3 as shown in figure 5.3.

Several questions arise in this context: how much energy savings can be realized by allowing this type of cooperation to take place? What level of coordination among the cooperating nodes is needed? And how must the route selection be done to maximize the energy savings?

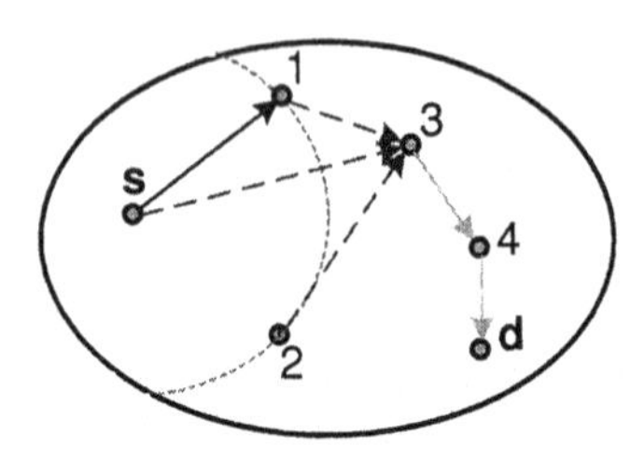

Figure 5.3. Cooperative Transmission

These are the problems that we look at here. We develop a formulation that captures the benefit of cooperative transmission and develop an algorithm for selecting the optimal route under this setting. We formulate the problem of finding the minimum energy cooperative route as two separate minimization problems. First, we look at the problem of optimal transmission of information between two sets of nodes. A separate problem is how to decide which nodes must be added to the reliable set in each transmission such that the information is routed to the final destination with minimum overall energy. We use dynamic programming to solve this second minimization problem. We present analytical results for the lower-bound of savings in networks with regular line or grid topology. We also propose two heuristics for finding the optimal path in arbitrary networks and present simulation results for the average energy savings of those heuristics.

2. Cooperative Transmission

Consider a wireless ad-hoc network consisting of arbitrarily distributed nodes where each node has a single omni-directional antenna. We assume that each node can dynamically adjust its transmitted power to control its transmission radius. It is also assumed that multiple nodes cooperating in sending the information to a single receiver node can precisely delay their transmitted signal to achieve perfect phase synchronization at the receiver. Under this setting, the information is routed from the source node to the destination node in a sequence of transmission slots, where each transmission slot corresponds to one use of the wireless medium. In each transmission slot/stage, either a node is selected to broadcast the information to a group of nodes or a subset of nodes that have already received the information cooperate to transmit that information to another group of nodes. As explained shortly, under our assumption it is only reasonable to restrict the size of the receiving set to one node when multiple nodes are cooperating in the transmission. So, each transmission is either a broadcast, where a single node is transmitting the information and the information is received by multiple nodes, or a cooperative, where multiple node simultaneously send the information to a single receiver. We refer to the first case as the *Broadcast Mode* and the second case at the *Cooperative Mode.* In the *Broadcast Mode*, we take advantage of the known *Wireless Broadcast Advantage.* In the *Cooperative Mode*, we benefit from the newly introduced concept of *Wireless Cooperative Advantage.*

The routing problem can be viewed as a multi-stage decision problem, where at each stage the decision is to pick the transmitting and the receiving set of nodes as well as the transmission power levels among all nodes transmitting in that stage. The objective is to get the information to the destination with minimum energy. The set of nodes that have the information at the kth stage is referred to as the kth-stage *Reliable Set*, $\mathsf{S_k}$, and the routing solution may be expressed as a sequence of expanding reliable sets that starts with only the source node and terminates as soon as the reliable set contains the destination node. We denote the transmitting set by S and the receiving set by T. The link cost between S and T, $\mathsf{LC(S,T)}$, is the minimum power needed for transmitting from S to T.

In this chapter, we make several idealized assumptions about the physical layer model. The wireless channel between any transmitting node, labeled $\mathsf{s_i}$, and any receiving node, labeled $\mathsf{t_j}$, is modeled by two parameters, its magnitude attenuation factor α_{ij} and its phase delay θ_{ij}. We assume that the channel parameters are estimated by the receiver and fed back to the transmitter. This assumption is reasonable for slowly varying channels, where the channel coherence time is much longer than the block transmission time. We also assume a free space propagation model where the power attenuation α_{ij}^2 is proportional to the inverse of the square of the distance between the communicating nodes $\mathsf{s_i}$ and $\mathsf{t_j}$. For the receiver model, we assume that the desired minimum transmission rate at the physical layer is fixed and nodes can only decode based on the signal energy collected in a single channel use. We also assume that the received information can be decoded with no errors if the received Signal-to-Noise ration, SNR, level is above a minimum threshold $\mathsf{SNR_{min}}$, and that no information is received otherwise. Without loss of generality, we assume that the information is encoded in a signal $\phi(\mathsf{t})$ that has unit power $\mathsf{P}_\phi = 1$ and that we are able to control the phase and magnitude of the signal arbitrarily by multiplying it by a complex scaling factor $\mathsf{w_i}$ before transmission. The transmitted power by node i is $|\mathsf{w_i}|^2$. The noise at the receiver is assumed to be additive, and the noise signal and power are denoted by $\eta(\mathsf{t})$ and P_η, respectively. This simple model allows us to find analytical results for achievable energy savings in some simple network topologies.

2.1 Link Cost Formulation

In this section, our objective is to understand the basic problem of optimal power allocation required for successful transmission of the same information from a set of source nodes $S = \{s_1, s_2, \cdots, s_n\}$ to a set of target nodes $T = \{t_1, t_2, \cdots, t_m\}$. In order to derive expressions for the link costs, we consider 4 distinct cases:

1 *Point-to-Point Link:* $n = 1, m = 1$: In this case, only one node is transmitting within a time slot to a single target node.

2 *Point-to-Multi-Point, Broadcast Link:* $n = 1, m > 1$: This type of link corresponds to the broadcast mode introduced in the last section. In this case, a single node is transmitting to multiple target nodes.

3 *Multi-Point-to-Point, Cooperative Link:* $n > 1, m = 1$: This type of link corresponds to the cooperative mode introduced in the last section. In this case, multiple nodes cooperate to transmit the same information to a single receiver node. We will assume that coherent reception, i.e. the transmitters are able to adjust their phases so that all signals arrive in phase at the receiver. In this case, the signals simply add up at the receiver and complete decoding is possible as long as the received SNR is above the minimum threshold SNR_{min}. Here, we do not address the feasibility of precise phase synchronization. The reader is referred to [9] for a discussion of mechanisms for achieving this level of synchronization.

4 *Multi-Point-to-Multi-Point Link:* $n > 1, m > 1$: This is not a valid option under our assumptions, as synchronizing transmissions for coherent reception at multiple receivers is not feasible. Therefore, we will not be considering this case.

***Point-to-Point Link:* $\mathbf{n = 1, m = 1}$.** In this case, $S = \{s_1\}$ and $T = \{t_1\}$. The channel parameters may be simply denoted by α and θ, and the transmitted signal is controlled through the scaling factor w. Although in general the scaling factor is a complex value, absorbing both power and phase adjustment by the transmitter, in this case we can ignore the phase as there is only a single receiver. The model assumptions made in Section 2 imply that the received signal is simply:

$$r(t) = \alpha e^{j\theta} w \phi(t) + \eta(t)\cdot$$

where $\phi(t)$ is the unit-power transmitted signal and $\eta(t)$ is the receiver noise with power P_η. The total transmitted power is $P_T = |w|^2$ and the SNR ratio at the receiver is $\frac{\alpha^2|w|^2}{P_\eta}$. For complete decoding at the receiver, the SNR must be above the threshold value SNR_{min}. Therefore the minimum power required, $\hat{P_T}$, and hence the point-to-point link cost $LC(s_1, t_1)$, is given by:

$$LC(s_1, t_1) \equiv \hat{P_T} = \frac{\mathsf{SNR}_{min} P_\eta}{\alpha^2}. \tag{5.1}$$

In equation 5.1, the point-to-point link cost is proportional to $\frac{1}{\alpha^2}$, which is the power attenuation in the wireless channel between s_1 and t_1, and therefore is proportional to the square of the distance between s_1 and t_1 under our propagation model.

Point-to-Multi-Point, Broadcast Link: $\mathbf{n = 1, m > 1.}$ In this case, $S = \{s_1\}$ and $T = \{t_1, t_2, \cdots, t_m\}$, hence m simultaneous SNR constraints must be satisfied at the receivers. Assuming that omni-directional antennas are being used, the signal transmitted by node s_1 is received by all nodes within a transmission radius proportional to the transmission power. Hence, a broadcast link can be treated as a set of point-to-point links and the cost of reaching a set of node is the maximum over the costs for reaching each of the nodes in the target set. Thus the minimum power required for the broadcast transmission, denoted by $LC(s_1, T)$, is given by:

$$LC(s, T) = \max\{LC(s_1, t_1), LC(s_1, t_2), \cdots, LC(s_1, t_m)\}. \tag{5.2}$$

Multi-Point-to-Point, Cooperative Link: $\mathbf{n > 1, m = 1.}$ In this case $S = \{s_1, s_2, \cdots, s_n\}$ and $T = \{t_1\}$. We assume that the n transmitters are able to adjust their phases in such a way that the signal at the receiver is:

$$r(t) = \sum_i^n \alpha_{i1}|w_i|\phi(t) + \eta(t).$$

The total transmitted power is $\sum_{i=1}^n |w_i|^2$ and the received signal power is $|\sum_{i=1}^n w_i\alpha_{i1}|^2$. The power allocation problem for this case is simply

$$\begin{aligned} \min \quad & \sum_{i=1}^n |w_i|^2 \\ \text{s.t.} \quad & \frac{|\sum_{i=1}^n w_i\alpha_{i1}|^2}{P_\eta} \geq \mathsf{SNR}_{min}. \end{aligned} \tag{5.3}$$

Lagrangian multiplier techniques may be used to solve the constrained optimization problem above. The resulting optimal allocation for each node i is given by

$$|\hat{w}_i| = \frac{\alpha_{i1}}{\sum_i^n \alpha_{i1}^2} \sqrt{SNR_{min} P_\eta}. \tag{5.4}$$

The resulting cooperative link cost $LC(S, t_1)$, defined as the optimal total power, is therefore given by

$$\begin{aligned} LC(S,t_1) &= \hat{P}_T \\ &= \sum_{i=1}^{n} |\hat{w}_i|^2 \\ &= \frac{1}{\sum_{i=1}^{n} \frac{\alpha_{i1}^2}{SNR_{min} P_\eta}}. \end{aligned} \tag{5.5}$$

It is easy to see that it can be written in terms of the point-to-point link costs between all the source nodes and the target nodes (see Equation 5.1) as follows:

$$LC(S, t_1) = \frac{1}{\frac{1}{LC(s_1,t_1)} + \frac{1}{LC(s_2,t_2)} + \cdots + \frac{1}{LC(s_n,t_1)}}. \tag{5.6}$$

A few observations are worth mentioning here. First, based on equation 5.4, the transmitted signal level is proportional to the channel attenuation. Therefore, in the cooperative mode *all* nodes in the reliable set cooperate to send the information to a single receiver. In addition, based on equation 5.6, the cooperative cost is smaller than each point-to-point cost. This conclusion is intuitively plausible and is a proof on the energy saving due to the *Wireless Cooperative Advantage*.

2.2 Optimal Cooperative Route Selection

The problem of finding the optimal cooperative route from the source node s to the destination node d, formulated in Section 2, can be mapped to a Dynamic Programming (DP) problem. The state of the system at stage k is the reliable set S_k, i.e. the set of nodes that have completely received the information by the kth transmission slot. The initial state S_0 is simply $\{s\}$, and the termination states are all sets that contain d. The decision variable at the kth stage is U_k, the set of nodes that will be added to the reliable set in the next transmission slot. The dynamical system evolves as follows:

$$S_{k+1} = S_k \cup U_k \quad k = 1, 2, \cdots \tag{5.7}$$

The objective is to find a sequence $\{U_k\}$ or alternatively $\{S_k\}$ so as to minimize the total transmitted power P_T, where

$$P_T = \sum_k LC(S_k, U_k) = \sum_k LC(S_k, S_{k+1} - S_k). \tag{5.8}$$

We will refer to the solution to this problem as the optimal transmission policy. The optimal transmission policy can be mapped to finding the shortest path in the state space of this dynamical system. The state space can be represented by as graph with all possible states, i.e. all possible subsets of nodes in the network, as its nodes. We refer to this graph as the *Cooperation Graph*. Figure 5.6 show the cooperation graph corresponding to the 4-node network shown in Figure 5.1.

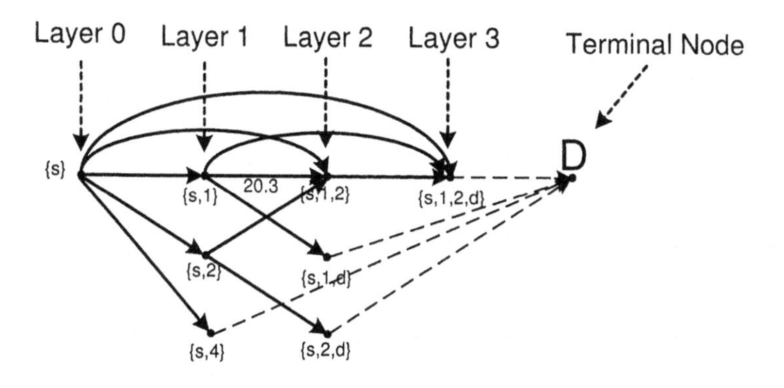

Figure 5.4. Cooperation Graph for a 4-Node Network

Nodes in the cooperation graph are connected with arcs representing the possible transitions between states. As the network nodes are allowed only to either fully cooperate or broadcast, the graph has a special layered structure as illustrated by Figure 5.6. All nodes in the kth layer are of size $k+1$, and a network with $n+1$ nodes the cooperation graph has n layers, and the kth layer has $\binom{n}{k}$ nodes. Arcs between nodes in adjacent layers correspond to cooperative links, whereas broadcast links are shown by cross-layer arcs. The costs on the arcs are the link costs defined in Section 2.1. All terminal states are connected to a single artificial terminal state, denoted by D, by a zero-cost arc. The optimal transmission policy is simply the shortest path between nodes s and D. There are 2^n nodes in the cooperation graph for a network with $n+1$ nodes. Therefore standard shortest path algorithms will in general have a complexity of $O(2^{2n})$. However, by taking advantage of some special properties of the cooperation graph, we are able to come up with an algorithm with complexity reduced to $O(n2^n)$. This algorithm is based on scanning the cooperation graph from left to right and constructing the shortest path to each nodes at the kth layer based on the shortest path to nodes in the previous layers. The *Sequential Scanning Algorithm* is outlined below.

Sequential Scanning Algorithm This is the algorithm for finding the optimal cooperative route in an arbitrary network based on finding the shortest path in the corresponding cooperation graph.

Initialize Initialize the cooperation graph data structure. Initialize the layer counter k to $k = 1$.

Repeat Construct to the shortest path to all nodes at the kth layer based on the shortest path to all nodes in the previous layers. Increment the counter.

Stop Stop when D is reached. i.e. when $k = n + 1$.

For a network with $n + 1$ nodes, the main loop in this algorithm is repeated n times and at the kth stage the shortest path to $\binom{n}{k}$ nodes must be calculated. This operation has a complexity of order $O(2^n)$, hence finding the optimal route is of complexity $O(n2^n)$.

Although the *Sequential Scanning Algorithm* substantially reduces the complexity for finding the optimal cooperative route in an arbitrary network, its complexity is still exponential in the number of nodes in the wireless network. For this reason, finding the optimal cooperative route in an arbitrary network becomes computationally intractable for larger networks. We will focus on developing computationally simpler and relatively efficient heuristics and on assessing their performance through simulation.

2.3 Example

Having developed the necessary mathematical tools, we now present a simple example that illustrates the benefit of cooperative routing. Figure 5.5 shows a simple network with 4 nodes. The arcs represent links and the arc labels are point-to-point link costs. The diagrams below show the six possible routes, P_0 through P_5. P_0 corresponds to a simple 2-hop, non-cooperative minimum energy path between s and d. P_1, P_2, and P_3 are 2-hop cooperative routes, whereas P_4 and P_5 are 3-hop cooperative routes. Figure 5.6 shows the corresponding cooperation graph for this network. Each transmission policy corresponds to a distinct path between $\{s\}$ and D in this graph and the minimum energy policy of P_3 corresponds to the shortest path. Table 5.1 lists the costs of the six policies.

3. Analytical Results for Line and Grid Topologies

In this section, we develop analytical results for achievable energy savings in line and grid networks. In particular, we consider a *Regular Line* Topology (see Figure 5.7) and a *Regular Grid* Topology (see Figure 5.8) where nodes are equi-distant from each other. Before proceeding further, let us define precisely

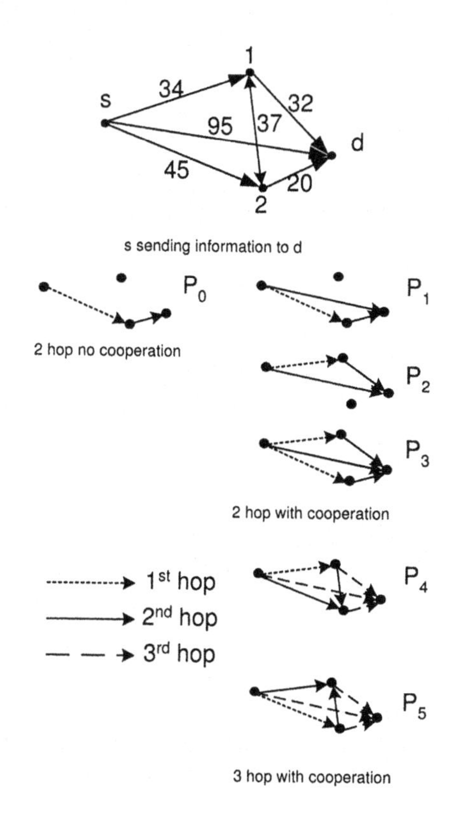

Figure 5.5. 4-Node Network Example

Figure 5.6. 4-Node Cooperation Graph

No.	Policy	Cost
P_0	$NonCooperative$	65
P_1	$(\{s\},\{s,2\},\{s,2,d\})$	≈ 61.5
P_2	$(\{s\},\{s,1\},\{s,1,d\})$	≈ 57.9
P_3	$(\{s\},\{s,1,2\},\{s,1,2,d\})$	≈ 55.9
P_4	$(\{s\},\{s,2\},\{s,1,2\},\{s,1,2,d\})$	≈ 73.6
P_5	$(\{s\},\{s,1\},\{s,1,2\},\{s,1,2,d\})$	≈ 65.2

Table 5.1. Transmission Policies for Figure 5.5

what we mean by energy savings for a cooperative routing strategy relative to the optimal non-cooperative strategy:

$$\mathsf{Savings} = \frac{\mathsf{P_T(Non-cooperative)} - \mathsf{P_T(Cooperative)}}{\mathsf{P_T(Non-cooperative)}}. \tag{5.9}$$

where $\mathsf{P_T(strategy)}$ denotes the total transmission power for the strategy.

3.1 Line Network-Analysis

Figure 5.7) shows a regular line where nodes are located at unit distance from each other on a straight line. In our proposed scheme, we restrict the cooperation to nodes along the optimal non-cooperative route. That is, at each transmission slot, all nodes that have received the information cooperate to send the information to the next node along the minimum energy non-cooperative route. This cooperation strategy is referred to as the CAN (*Cooperation Along the Minimum Energy Non-Cooperative Path*) strategy.

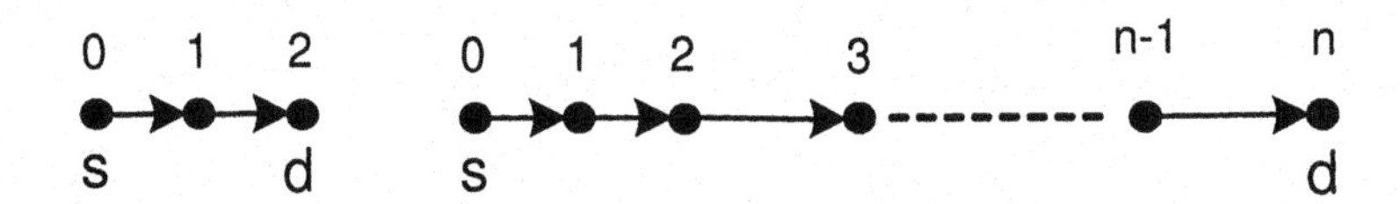

Figure 5.7. Regular Line Topology

For the 3-node line network in Figure 5.7, it is easy to show that the optimal non-cooperative routing strategy is to relay the information through the middle node. Since a longer line network can be broken down into short 2-hop components, it is clear that the optimal non-cooperative routing strategy is to always send the information to the next nearest node in the direction of the destination until the destination node is reached. From Equation 5.1, the link cost for every stage is $\frac{\mathsf{SNR_{min}P_\eta}}{\alpha^2}$, where α is the magnitude attenuation between two adjacent nodes 1-distance unit apart. Under our assumptions, α^2 is proportional to the inverse of the distance squared. Therefore,

$$P_T(\text{Non}-\text{cooperative}) = n\frac{SNR_{min}P_\eta}{\alpha^2}. \tag{5.10}$$

With the CAN strategy, after the mth transmission slot, the reliable set is $S_m = \{s, 1, \cdots, m\}$, and the link cost associated with the nodes in S_m cooperating to send the information to the next node $(m+1)$ follows from Equation 5.6 and is given by

$$LC(S_m, m+1) = \frac{SNR_{min}P_\eta}{\sum_{i=1}^{m+1}\frac{\alpha^2}{i^2}}. \tag{5.11}$$

Therefore, the total transmission power for the CAN strategy is

$$\begin{aligned} P_T(CAN) &= \sum_{m=0}^{n-1} LC(S_m, m+1) \\ &= \frac{SNR_{min}P_\eta}{\alpha^2}\sum_{m=0}^{n-1}\frac{1}{C(m+1)}, \end{aligned} \tag{5.12}$$

$$\text{where } C(m) = \sum_{i=1}^{m}\frac{1}{i^2}. \tag{5.13}$$

Before moving to find the savings achieved by *CAN* in a line, we need to proves the following simple lemma regarding the existence of the average of terms for a decreasing sequence.

LEMMA 5.1 *Let* a_n *be a decreasing sequence with a finite limit* c, *then:* $\lim_{m\to\infty}\frac{\sum_{n=1}^{m}a_n}{m} = c$.

Proof: For any value of m, let m_0 be an arbitrary integer less than m:

$$\begin{aligned} \lim_{m\to\infty}\frac{\sum_{n=1}^{m}a_n}{m} &= \lim_{m\to\infty}\frac{1}{m}\left(\sum_{n=1}^{m_0}a_n + \sum_{n=m_0+1}^{m}a_n\right) \\ &= \lim_{m\to\infty}\frac{1}{m}\sum_{n=1}^{m_0}a_n + \lim_{m\to\infty}\frac{1}{m}\sum_{n=m_0+1}^{m}a_n \\ &= 0 + \lim_{m\to\infty}\frac{1}{m}\frac{m-(m_0+1)}{m-(m_0+1)}\sum_{n=m_0+1}^{m}a_n \\ &= \lim_{m\to\infty}\frac{m-(m_0+1)}{m}\frac{1}{m-(m_0+1)}\sum_{n=m_0+1}^{m}a_n \\ &= \lim_{m\to\infty}\frac{m-(m_0+1)}{m}\lim_{m\to\infty}\frac{1}{m-(m_0+1)}\sum_{n=m_0+1}^{m}a_n \end{aligned}$$

$$= \lim_{m\to\infty} \frac{1}{m-(m_0+1)} \sum_{n=m_0+1}^{m} a_n \,. \tag{5.14}$$

Since a_n is a decreasing sequence, all terms in the final sum are less than a_{m_0}. Furthermore, $\lim_{n\to\infty} a_n = c$. So, all terms in the final sum are greater than c. Hence:

$$c \le \lim_{m\to\infty} \frac{\sum_{n=1}^{m} a_n}{m} = \lim_{m\to\infty} \frac{1}{m-(m_0+1)} \sum_{n=m_0+1}^{m} a_n \le a_{m_0}. \tag{5.15}$$

For increasing values of m, m_0 may be chosen such that a_{m_0} is arbitrarily close to c and the proof is established.

THEOREM 5.2 *For a regular line network as shown in Figure 5.7, the CAN strategy results in energy savings of* $(1 - \frac{1}{n}\sum_{m=1}^{n}\frac{1}{C(m)})$. *As the number of nodes in the network grows, the energy savings value approaches* $(1 - \frac{6}{\pi^2}) \approx 39\%$.

Proof: The minimum energy non-cooperative routing a regular line network with n hops has cost equal to n. The cost of the optimal cooperation scheme, i.e. the CAN strategy, is:

$$P_T(\text{Cooperative}) = \sum_{m=1}^{n} LC(\{s,\cdots,m-1\},m) = \sum_{m=1}^{n} \frac{1}{C(m)} \tag{5.16}$$

where $C(m)$ is defined by equation 5.13. The energy savings achieved, as defined by equation 5.11, is:

$$\text{Savings}(n) = \frac{P_T(\text{Non}-\text{Cooperative}) - P_T(\text{Cooperative})}{P_T(\text{Non}-\text{Cooperative})} \tag{5.17}$$

$$= \frac{n - \sum_{m=1}^{n}\frac{1}{C(m)}}{n} \tag{5.18}$$

$$= 1 - \frac{1}{n}\sum_{m=1}^{n}\frac{1}{C(m)} \tag{5.19}$$

$\frac{1}{C(m)}$ is a decreasing sequence with limit of $\frac{6}{\pi^2}$. So, based on lemma 5.1 we have:

$$\lim_{n\to\infty} \text{Savings}(n) = 1 - \lim_{n\to\infty} \frac{1}{n}\sum_{m=1}^{n}\frac{1}{C(m)} = 1 - \frac{6}{\pi^2} \tag{5.20}$$

This establishes the claim and completes the proof.

3.2 Grid Network

Figure 5.8 shows a regular $n \times n$ grid topology with s and d located at opposite corners. A $n \times n$ grid can be decomposed into many 2×2 grid. Assuming that the nodes are located at a unit distance from each other, in a 2×2 grid, a diagonal transmission has a cost of 2 units, equal to the cost of one horizontal and one vertical transmission. For this reason, in an $n \times n$ grid there are many non-cooperative routes with equal cost. Figure 5.8 shows two such routes for an $n \times n$ grid.

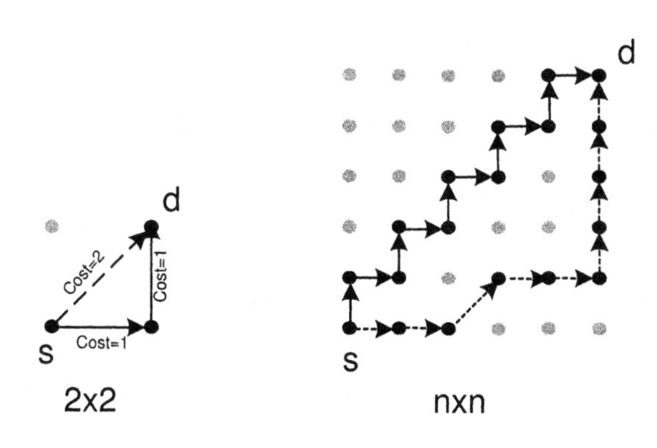

Figure 5.8. Regular Grid Topology

The minimum-energy non-cooperative route is obtained by a stair-like policy (illustrated in Figure 5.8), and its total power is $2n$. We will base our analysis for deriving the bound for saving based on this stair-like non-cooperative path. The following theorem stated the energy savings achieved by the CAN strategy applied to this non-cooperative route.

THEOREM 5.3 *For a regular grid network as shown in Figure 5.8, the energy savings achieved by using the CAN strategy approaches* 56% *for large networks.*

Proof: Figure 5.9 shows an intermediate step in routing the information is a regular grid. At this stage, all the nodes with a darker shade, nodes 1 through 8, have received the information. In the next step, the information must be relayed to node 9. The cooperative cost of this stage is

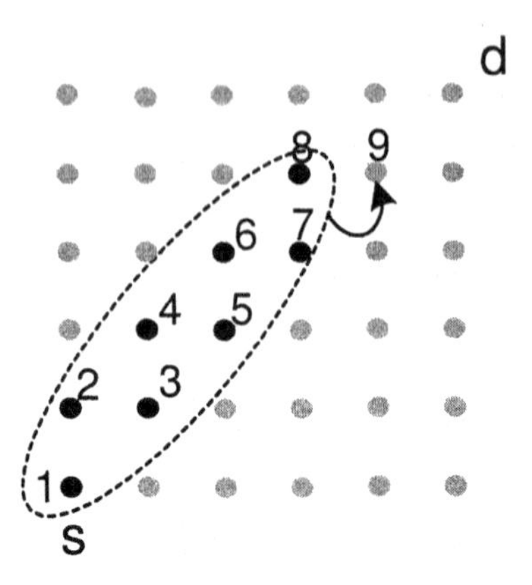

Figure 5.9. Cooperative Routing in a Grid Topology

$$\begin{aligned} LC(\{1,\cdots,8\},9) &= \frac{1}{\sum_{i=1}^{8}\frac{1}{LC(i,9)}} \\ &= \frac{1}{\frac{1}{1}+\frac{1}{2}+\frac{1}{5}+\frac{1}{8}+\frac{1}{13}+\frac{1}{18}+\frac{1}{25}+\frac{1}{32}} \qquad (5.21) \\ &= \frac{1}{\underbrace{\frac{1}{1}+\frac{1}{5}+\frac{1}{13}+\frac{1}{25}}+\underbrace{\frac{1}{2}+\frac{1}{8}+\frac{1}{18}+\frac{1}{32}}} \qquad (5.22) \end{aligned}$$

In general, the cooperative cost of the mth stage of the proposed strategy is

$$\begin{aligned} C_{grid}(m) &= LC(\{1,\cdots,m\},m+1) \\ &= \frac{1}{\sum_{i=1}^{m}\frac{1}{LC(i,m)}} \end{aligned} \qquad (5.23)$$

It is not too hard to see that the point-to-point costs have the following form

$$LC(i,m) = \left(\left\lceil\frac{m-i}{2}\right\rceil\right)^2 + \left(\left\lfloor\frac{m-i}{2}\right\rfloor\right)^2 \qquad (5.24)$$

Using Equation 5.24, Equation 5.23 can be written as

$$\begin{aligned} C_{grid}(m) &= \frac{1}{\sum_{i=1}^{m}\frac{1}{LC(i,m)}} \\ &= \frac{1}{\sum_{i=1}^{m}\frac{1}{(\lceil\frac{m-i}{2}\rceil)^2+(\lfloor\frac{m-i}{2}\rfloor)^2}} \end{aligned}$$

$$= \frac{1}{\sum_{k=1}^{\lceil \frac{m}{2} \rceil} \frac{1}{2k^2-2k+1} + \sum_{k=1}^{\lfloor \frac{m}{2} \rfloor} \frac{1}{2k^2}} \tag{5.25}$$

Comparing Equation 5.22 and Equation 5.25, it is easy to see that the first group of terms is generated by the first sum term and the second group is generated by the second sum term. $C_{grid}(m)$ is a decreasing sequence of numbers and can be shown, using Maple, to have a limit equal to 0.44.

The total cost for the cooperative route in an $n \times n$ grid is

$$P_T(\text{Cooperative}) = \sum_{m=1}^{2n} C_{grid}(m) \tag{5.26}$$

The energy saving, as defined by equation 5.9, is

$$\begin{aligned} \text{Savings}(n) &= \frac{P_T(\text{Non}-\text{Cooperative}) - P_T(\text{Cooperative})}{P_T(\text{Non}-\text{Cooperative})} \\ &= \frac{2n - \sum_{m=1}^{2n} C_{grid}(m)}{2n} \\ &= 1 - \frac{1}{2n}\sum_{m=1}^{2n} C_{grid}(m) \end{aligned} \tag{5.27}$$

Since $C_{grid}(m)$ is a decreasing sequence and $\lim_{m\to\infty} C_{grid}(m) = 0.44$, by lemma 5.1, the savings in the case of a regular grid, as calculated in equation 5.27, approaches $1 - 0.44 = 56\%$. This establishes the claim and completes the proof for the lower bound of achievable savings in a regular grid.

4. Heuristics & Simulation Results

We present two possible general heuristic schemes and related simulation results. The simulations are over a network generated by randomly placing nodes on an 100×100 grid and randomly choosing a pair of nodes to be the source and destination. For each realization, the minimum energy non-cooperative path was found. Also, the proposed heuristic were used to find co-operative paths. The performance results reported are the energy savings of the resulting strategy with respect to the optimal non-cooperative path averaged over $100,000$ simulation runs.

The two heuristics analyzed are outlined below.

CAN-L Heuristic *Cooperation* **A***long the* **N***on-Cooperative Optimal Route*:

This heuristic is based on the CAN strategy described Section 3. CAN-L is a variant of CAN as it limits the number of nodes allowed to participate in the cooperative transmission to L. In particular, these nodes are chosen to be the last L nodes along the minimum energy non-cooperative path. As mentioned before, in each step the last L nodes cooperate to transmit the information to the next node along the optimal non-cooperative path. The only processing needed in this class of algorithm is to find the optimal non-cooperative route. For this reason, the complexity of this class of algorithms is the same as finding the optimal non-cooperative path in a network or $O(N^2)$.

PC-L Heuristic P*rogressive* C*ooperation*:

Initialize Initialize *B*est Path to the optimal non-cooperative route. Initialize the *S*uper Node to contain only the source node.

Repeat Send the information to the first node along the current *B*est Path. Update the *S*uper Node to include all past L nodes along the current *B*est Path. Update the link costs accordingly, i.e. by considering the *S*uper Node as a single node and by using equation 5.6. Compute the optimal non-cooperative route for the new network/graph and update the *B*est Path accordingly.

Stop Stop as soon as the destination node receives the information.

For example, with $L = 3$, this algorithm always combines the last 3 nodes along the current *B*est Route into a single node, finds the shortest path from that combined node to the destination and send the information to the next node along that route. This algorithm turns out to have a complexity of $O(N^3)$ since the main loop is repeated $O(N)$ times and each repetition has a complexity of $O(N^2)$.

A variant of this algorithm keeps a window W of the most recent nodes, and in each step all subsets of size L among the last W nodes are examined and the path with the least cost is chosen. This variant has a complexity of $O\left(\binom{W}{L} \times N^3\right)$, where W is the window size. We refer to this variant as *Progressive Cooperation with Window*.

Figures 5.10 and 5.11 show average energy savings ranging from 20% to 50% for CAN and PC algorithms. It can be seen that PC-2 performs almost as well as CAN-3 and PC-3 performs much between than CAN-4. This show that the method for approximating the optimal route is very important factor in increasing the savings. Figures 5.12 compares CAN, PC, and PC-W on the same chart. It is seen that PC-3-4 performs better than PC-3, which performs substantially better than CAN-4. In general, it can be seen that the energy savings increase with L, and that improvements in savings are smaller for larger values of L. As there is a trade-off between the algorithm complexity and the algorithm performance, these simulation results indicate that it would be reasonable to chose L to be around 3 or 4 for both the CAN and PC heuristics.

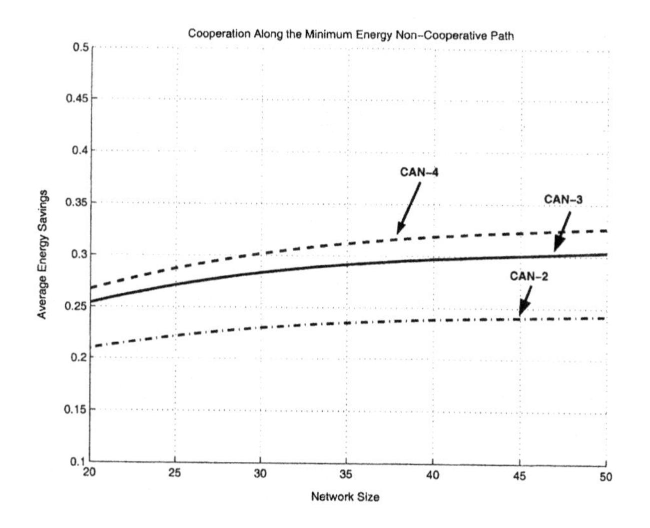

Figure 5.10. Performance of CAN

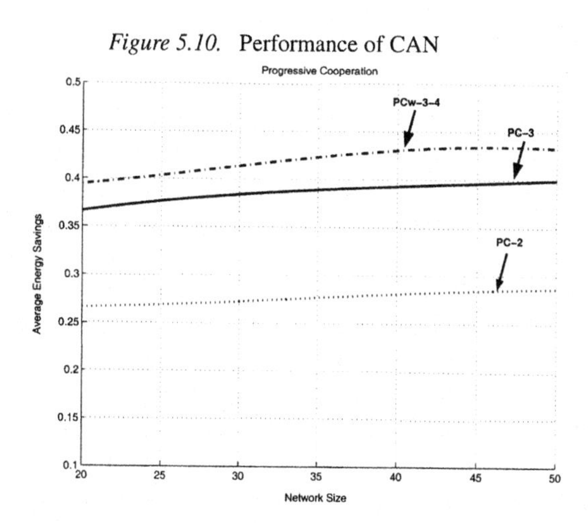

Figure 5.11. Performance of PC

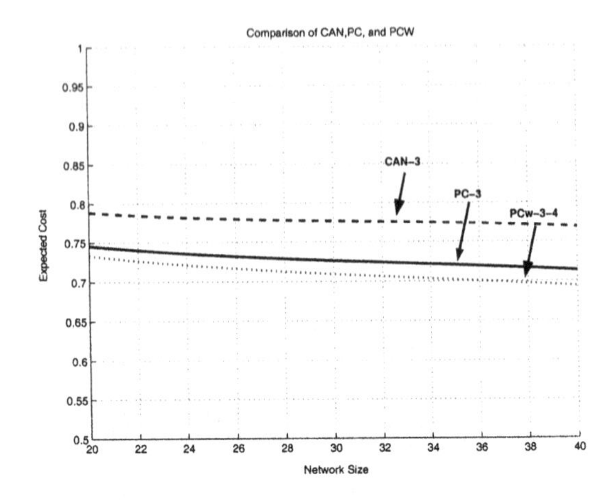

Figure 5.12. Comparison

5. Conclusions

In this chapter we formulated the problem of finding the minimum energy cooperative route for a wireless network under idealized channel and receiver models. Our main assumption were that the channel states are known at the transmitter and precise power and phase control, to achieve coherent reception is possible. We focused on the optimal transmission of a single message from a source to destination through sets of nodes, that may act as cooperating relays. Fundamental to the understanding of the routing problem was the understanding of the optimal power allocation for a single message transmission from a set of source nodes to a set of destination nodes. We presented solutions to this problem, and used these as the basis for solving the minimum energy cooperative routing problem. We used Dynamic Programming (DP) to formulate the cooperative routing problem as a multi-stage decision problem. However, general shortest algorithms are not computationally tractable and are not appropriate for large networks. For a Regular Grid Topology and a Regular Grid Topology, we analytically obtained the energy savings due to cooperative transmission, demonstrating the benefits of the proposed cooperative routing scheme. For general topologies, we proposed two heuristics and showed significant energy savings (close to 50%) based on simulation results.

References

[1] L.M. Feeney, M. Nilsson, "Investigating the energy consumption of a wireless network interface in an ad hoc networking environment," INFOCOM 2001, pp. 1548 -1557

[2] R.T. Derryberry, S.D. Gray, D.M. Ionescu, G. Mandyam, B. Raghothaman, "Transmit diversity in 3G CDMA systems," IEEE Communications Magazine, April 2002, pp. 68-75

[3] J.E. Wieselthier, G.D. Nguyen, A. Ephremides, "Algorithms for energy-efficient multicasting in ad hoc wireless networks," MILCOM 1999, pp. 1414-1418

[4] J.E. Wieselthier, G.D. Nguyen, A. Ephremides, "On the construction of energy-efficient broadcast and multicast trees in wireless networks ," INFOCOM 2000, pp. 585 -594

[5] Ahluwalia Ashwinder, Eytan Modiano and Li Shu, "On the Complexity and Distributed Construction of Energy Efficient Broadcast Trees in Static Ad Hoc Wireless Networks," Conference on Information Science and System, Princeton, NJ, March, 2002

[6] Mario Cagalj, Jean-Pierre Hubaux and Christian Enz,"Minimum-energy broadcast in all-wireless networks: NP-completeness and distribution issues," ACM MobiCom 2002

[7] Ivana Maric, Roy Yates, "Efficient Multihop Broadcast for Wideband Systems", DIMACS Workshop on Signal Processing for Wireless Transmission Rutgers University, Piscataway, NJ, October 7-9, 2002

[8] Anand Srinivas and Eytan Modiano, "Minimum Energy Disjoint Path Routing in Wireless Ad Hoc Networks," ACM Mobicom 2003

[9] T. Yung-Szu , G.J. Pottie, "Coherent cooperative transmission from multiple adjacent antennas to a distant stationary antenna through AWGN channels," IEEE Vehicular Technology Conference, 2002, pp. 130-134 vol.1

Chapter 6

IN-NETWORK DECISION MAKING VIA LOCAL MESSAGE-PASSING*

Murat Alanyali and Venkatesh Saligrama
Department of Electrical and Computer Engineering
Boston University, Boston, MA 02215
{alanyali,srv}@bu.edu

Abstract We consider in-network data processing to classify an unknown event based on noisy sensor measurements. The sensors are distributed and can only exchange messages through a network. The sensor network is modeled by means of a graph, which captures the connectivity of different sensor nodes in the network. The task is to arrive at a consensus about the event after exchanging such messages. The focus of this paper is twofold: a) characterize conditions for reaching a consensus; b) derive conditions for when the consensus converges to the centralized MAP estimate. The novelty of the paper lies in applying belief propagation as a message passing strategy to solve a distributed hypothesis testing problem for a pre-specified network connectivity. We show that the message evolution can be re-formulated as the evolution of a linear dynamical system, which is primarily characterized by network connectivity. This leads to a fundamental understanding of as to which network topologies naturally lend themselves to consensus building and conflict avoidance.

Keywords: Sensor networks, in-network data processing, belief propagation, statistical decision making.

1. Introduction

Recent advances in sensor and computing technologies [11, 15, 8] enable massively distributed networks of sensors as candidate technologies to provide real-time information in diverse applications such as building safety, environ-

*This work was supported by NSF CAREER Program under grant ANI-0238397 and ONR Young Investigator Award N00014-02-100362.

mental remediation, habitat monitoring, power systems and manufacturing [13]. These networks are constructed of tiny and relatively unsophisticated devices that are capable of sensing, processing and exchanging data over a wireless medium, producing aggregate results far greater than their individual capabilities. The potential of networked sensing by such devices is enormous, yet there are significant challenges in realizing this potential by designs that are reliable, efficient and scalable.

The broad aim of networked sensing is to use a distributed group of sensors and decision agents to reliably classify, localize, and track relevant dynamic events in a timely manner within the constraints imposed by the ad hoc networking environment. While this task touches upon problems studied in the contexts of distributed computing and ad hoc networking, standard results from these areas alone are not adequate in the present setting due to the following factors: (i) Inherent uncertainty: sensors operate in a noisy environment and therefore generate unreliable data, which leads to local false alarms. In addition, the underlying process may display rapid, unanticipated change, leading to temporal uncertainty. Finally, many processes of interest display high spatial variability, yet may be only sparsely sampled in this dimension. (ii) Robustness: sensors are prone to failures, thereby robustness is an indispensable quality for proper operation. (iii) Ad-hoc networking environment: Envisioned applications require ad-hoc operating modes that do not admit stringent planning of the communication infrastructure. This restriction entails uncontrolled topologies, unscheduled sensor transmissions and a predominantly event-driven network operation. (iv) Power limitations: Transmit power determines communication range, and leads to an interplay between logical network topology and lifetime of a network of energy-limited sensors. Put another way, increased direct centralization bears an increased power cost, and thus raises a question of how to balance coordination against a finite energy source. The energy limitation also raises the question of *what* should be transmitted by individual sensors in order to achieve efficient overall system operation while conserving energy. This latter observation points to adapting and refining data by cooperating with other sensors as a critical asset in network efficiency.

In this paper we focus on a scenario involving noisy sensors observing a single, common event. The sensors operate in a distributed fashion in that they make local measurements and can only exchange messages through a network. The sensor network is thereby modeled by means of a graph, which captures the connectivity of different sensor nodes in the network. The task is to arrive at a certain consensus about the event after exchanging such messages. In principle the consensus is intended to classify the observed event among one of M hypotheses. If the observations are centrally available there is a well-established solution methodology for optimal solutions of such problems [33]. Fundamental questions arise when data is distributed and the centralized solutions are no longer feasible due to time and communication rate constraints.

Unlike the traditional settings our attention is limited to: a) fixed network topology that provides a fixed routing mechanism between different sensor nodes; b) exchanging *informative* data, as opposed to local decisions, between different sensor nodes. Pearl's belief propagation algorithm [26] is investigated here as a natural mechanism for exchanging informative data. The main idea can be explained as follows: A sensor node j sending messages to node k summarizes information received from all the other nodes it is connected to and forwards this information to node k. The messages are generally in the form of node k's conditional marginal distribution. Node k then updates its posterior probability, which is usually called as the belief. The process continues with node k updating its messages to be sent to its neighbors and so on.

The organization of the paper is as follows. Section 2 gives an overview of existing related work on the subject. In Section 3 we provide a description of the problem setup. In Section 4 belief propagation algorithm is discussed. Sections 5 and 6 provides the main result and numerical examples. Section 7 concludes with a discussion.

2. Related Work

Networked sensing has received significant attention within the networking, signal processing and information-theory communities.

The networking community (see [35, 32, 19] and references therein) has largely addressed the problem from the perspective of ad-hoc networks and routing. On the one hand, ad-hoc networking protocols offer the possibility of networked communication on amorphous topologies that arise due to wireless communications. Both routing [25, 16] and energy efficiency [1, 7, 28, 20, 29] have been vigorously investigated in this context; in turn ad-hoc networking techniques emerge as plausible candidates for the networking layer in sensor networks. On the other hand, these techniques may be questioned in their reliance on the layered protocol architecture of traditional data networking. In particular, as opposed to conventional data networks, sensor networks have a single, overriding goal that pertains to the broad task of inferencing. The layered architecture has significant operational consequences in this respect: It effectively separates sensing and networking, and thereby restricts the sensor network to query-based operation carried out by fusion centers. In addition to evident robustness issues, it is unclear if aggregating data at designated centers leads to an efficient solution for networked sensing. A conceptual alternative for the layered architecture is in-network data processing, which refers to progressive refinement of data within the network. Potential benefits of this alternative are strongly hinted by Kalman filtering in the context of control theory. There is a recent surge of interest in in-network data aggregation, and much of this activity focuses on distributed computation of several functions, such as the sum and the maximum, of the overall data [36–40].

Distributed detection has had a long history in control, signal processing and information theory [30, 10, 3, 5, 31, 34, 2, 23, 27, 14, 9, 18, 6, 21, 24]. Although, these approaches vary significantly, they share a common distributed computational viewpoint. The sensor nodes are organized into a simple information-flow architecture, such as: a) information flowing from each sensor to a unique fusion node; b) information collected locally at intermediate fusion nodes are transmitted and fused at the root node. In decentralized detection, the information flow for each sensor is a local decision rule which takes values on a finite alphabet (the aim here is to account for limited communications). The pertinent question then is to determine the local decision as well as fusion rules that optimizes misclassification. Distributed estimation of ergodic sources has been subjected to intense research recently in the information theory [2, 23, 27, 9, 18] literature. Again the information flow architecture is similar. Each sensor observes a spatially correlated ergodic source in white noise. The task is to determine the minimum mean-squared error achievable at the fusion center. Recently, similar ideas have been applied for boundary localization [21, 22] as well.

There are several drawbacks of these approaches. First, a distributed computational approach lacks flexibility in that it may be well suited for a particular inferencing task, while being unsuitable for other tasks. Second, the solution requires centralized design of decision and fusion strategies. This is not meaningful for the following reasons: (a) such a centralized design and decentralized implementation is generally computationally intractable [30]; (b) centralized design will require global knowledge of individual sensor models, which is not meaningful on account of significant uncertainties and variation that are inherent in the environment; (c) lack of robustness, i.e., how to modify the strategies in light of sensor failures is unclear. Indeed, the last issue is common in an ad-hoc networking environment, wherein packet losses from individual sensors happen quite often.

3. Distributed Classification Problem

We consider classifying an unknown event based on noisy observations. Namely, we denote by $\mathcal{H} = \{H_1, H_2, \cdots, H_M\}$ the possible classes that the observed event may belong to, and consider a Bayesian setting where π_o is the prior probability distribution on $\mathcal{H}$ before any measurements are taken. The measurements $(X_v : v \in V)$ are indexed by a set V of sensors, and each entry represents the value observed by a distinct sensor. For each $m = 1, 2, \cdots, M$, let $f_m : \mathbf{R}^V \mapsto \mathbf{R}_+$ be the conditional probability density function of $(X_v : v \in V)$ given that H_m is the true hypothesis. We shall assume that observations are conditionally independent given the true hypothesis. That is,

$$f_m(x) = \prod_{v \in V} f_m^v(x_v), \quad x = (x_v : v \in V) \in \mathbf{R}^V$$

for marginal densities $f_m^v : \mathbf{R} \mapsto \mathbf{R}_+$, $v \in V$. Given $X_v = x_v$ for $v \in V$, the posterior distribution π of the true hypothesis is then identified uniquely by the relation

$$\pi(H_m) \propto \pi_o(H_m) \prod_{v \in V} f_m^v(x_v), \quad m = 1, 2, \cdots, M. \tag{6.1}$$

Of particular interest here is the maximum a-posteriori probability (MAP) estimate of the true hypothesis given the observations $(X_v : v \in V)$. Namely, hypothesis H_{m^*} is a MAP estimate if

$$\pi_o(H_{m^*}) \prod_{v \in V} f_{m^*}^v(x_v) = \max_m \left\{ \pi_o(H_m) \prod_{v \in V} f_m^v(x_v) \right\}.$$

We concentrate on distributed applications in which a single decision maker that has access to all observations $(X_v : v \in V)$ is not available. Instead, it is assumed that each sensor can communicate with a certain subset of other sensors, and thereby forms an estimate of the posterior distribution π based on both its own observation and its prior correspondence with its neighbors. The objective of the paper is to identify communication schemes which guarantee that each sensor eventually identifies a MAP estimate. Furthermore applications of interest concern vast numbers of sensors; in turn non-scalable schemes such as simple flooding of observations are excluded from the present discussion. Specifically, we examine the performance of Pearl's belief propagation algorithm [26], which is subject to considerable recent interest in similar statistical inference problems that arise, for example, in coding and artificial intelligence [17, 26].

The communication structure among the sensors is represented via a directed graph $G = (V, E)$. The vertices V of this graph correspond to sensors, and an ordered pair (v', v) of vertices belongs to the edge set E if and only if there exists a communication link from sensor v' to sensor v. We will identify each edge $e \in E$ with its source vertex $s(e)$ and its destination vertex $d(e)$ so that $e = (s(e), d(e))$. Sensor v' is referred to as a *neighbor* of sensor v if there is a link from vertex v' to vertex v in G, so that sensor v' can send a message to sensor v. Let $N(v)$ denote the set of neighbors of sensor v so that

$$N(v) = \{v' \in V : (v', v) \in E\}, \quad v \in V.$$

The communication graph G is not required to bear any relationship to the underlying statistical model.

4. Belief Propagation

We start with a brief digression to statistical inferencing via belief propagation in order to motivate the distributed message passing algorithm adopted here. Let $(Y_v : v \in V)$ be a random vector with values in $\mathcal{Y}^V$, and for certain mappings $\phi_v : \mathcal{Y} \mapsto \mathbf{R}_+$, $v \in V$, and $\psi_e : \mathcal{Y}^2 \mapsto \mathbf{R}_+$, $e \in E$, let the distribution of $(Y_v : v \in V)$ satisfy

$$P(Y_v = y_v : v \in V) \propto \prod_{v \in V} \phi_v(y_v) \prod_{e \in E} \psi_e(y_{s(e)}, y_{d(e)}), \qquad (6.2)$$

for $y_v \in \mathcal{Y}$, $v \in V$. Such graphical models arise in a variety of contexts where efficient computation of marginal distributions $P(Y_v = y_v)$, $v \in V$, is of interest. Let the undirected graph $\tilde{G} = (V, \tilde{E})$ be defined so that the unordered pair $[v, v'] \in \tilde{E}$ if and only if $(v, v') \in E$ or $(v', v) \in E$. It is well-known that if $\tilde{G}$ is a tree, then local message passing via Pearl's sum-product algorithm [26] results in distributed, local computation of the marginal distributions. Namely, let the kth message sent from sensor $v' \in N(v)$ to sensor v be the vector $m_k^{(v',v)} = (m_k^{(v',v)}(y) : y \in \mathcal{Y})$ defined by

$$\begin{aligned} m_0^{(v',v)}(y) &= 1 \\ m_k^{(v',v)}(y) &= \sum_{y' \in \mathcal{Y}} \psi_{(v',v)}(y', y)\phi_{v'}(y') \prod_{\hat{v} \in N(v') - \{v\}} m_{k-1}^{(\hat{v},v')}(y'), \end{aligned}$$

$k \geq 1$, and suppose that upon receiving the kth messages from all of its neighbors, each sensor $v \in V$ constructs an estimate $\hat{\pi}_k^v$ of the local marginal distribution by setting

$$\hat{\pi}_k^v(y) \propto \phi_v(y) \prod_{v' \in N(v)} m_k^{(v',v)}(y), \quad y \in \mathcal{Y}. \qquad (6.3)$$

Then $\hat{\pi}_k^v(y)$ converges to the correct marginal distribution $P(Y_v = y)$ within a finite number of steps, provided that $\tilde{G}$ is a tree.

Consider next the sum-product algorithm with the following parameterization:

$$\begin{aligned} \mathcal{Y} &= \mathcal{H} \\ \phi_v(H_m) &= f_m^v(x_v) \sqrt[|V|]{\pi_o(H_m)}, \\ \psi_e(H_j, H_m) &= \mathbf{1}\{j = m\}, \quad j, m = 1, 2, \cdots, M, \end{aligned} \qquad (6.4)$$

where $\mathbf{1}\{\cdot\}$ denotes the indicator function whose value is 1 if its argument is correct and is 0 otherwise. It is straightforward to verify that equality (6.2) reduces to

$$P(Y_v = y_v : v \in V) \propto \mathbf{1}\{y_v = y_{v'} \text{ for all } v, v' \in V\}\pi(y_{v^*}),$$

where π is given by relation (6.1) and $v^* \in V$ is arbitrary. In particular $P(Y_v = Y_{v'} \text{ for all } v, v' \in V) = 1$ and each marginal Y_v has distribution π. The sum-product algorithm now prescribes the message composition

$$m_0^{(v',v)}(h) = 1 \tag{6.5}$$
$$m_k^{(v',v)}(h) = \phi_{v'}(h) \prod_{\hat{v} \in N(v')-\{v\}} m_{k-1}^{(\hat{v},v')}(h), \tag{6.6}$$

for each $h \in \{H_1, H_2, \cdots, H_m\}$, $k \geq 1$. From the prior discussion it is clear that if $\tilde{G}$ is a tree, then the algorithm assures that $\hat{\pi}_k^v = \pi$ for large enough k, hence each sensor can identify a MAP estimate based on the global observation set.

On the one hand, the message passing algorithm (6.5)–(6.6) has an evident practical appeal: Each message is determined locally by the observation at the sensor and the prior messages received from neighboring sensors. Furthermore, the algorithm entails a relaxed synchronization among sensors, as it can be implemented by programming each sensor to send out initial messages immediately and to send out its kth messages only after receiving $(k-1)$th messages from all of its neighbors. On the other hand, asymptotic features of the sum-product in general topologies of G are not well-understood. In fact, there is ample evidence that the algorithm may, in general, fail to converge, or may converge to an inaccurate estimate of the marginal distributions. We next address these issues in the particular instantiation that pertains to the classification problem, and give an account of the asymptotic behavior of the estimates $\hat{\pi}_k^v : v \in V$ for general graphs.

5. Main Result

For each pair of edges $e, e' \in E$ let

$$a_{e,e'} = \mathbf{1}\{d(e') = s(e),\ s(e') \neq d(e)\}.$$

Note that $a_{e,e'} = 1$ if and only if edge e' leads to the origin of edge e but the ordered pair (e', e) is not a directed cycle. For each hypothesis $h \in \{H_1, H_2, \cdots, H_m\}$ let

$$u^h(v) = \log(\phi_v(h)), \quad v \in V$$
$$x_k^h(e) = \log(m_k^e(h)), \quad e \in E.$$

Taking the logarithm of both sides in equalities (6.5)–(6.6) leads to the linear system

$$x_k^h(e) = u^h(s(e)) + \sum_{e' \in E} a_{e,e'} x_{k-1}^h(e'), \quad x_0^h(e) = 0. \tag{6.7}$$

Define the vector $u^h = (u^h(s(e)) : e \in E)$ and define the binary matrix $A = [a_{e,e'}]_{E\times E}$, so that equality (6.7) takes the vector form

$$x_k^h = u^h + Ax_{k-1}^h, \quad x_0^h = 0.$$

Note in particular that

$$x_k^h = \sum_{j=0}^{k-1} A^j u^h, \quad k \geq 1 \tag{6.8}$$

and $A^j = [a^j_{e,e'}]_{E\times E}$ where $a^j_{e,e'}$ is the number of directed paths of length j that start with edge e', end with edge e, and that do not have any 2-hop cycles.

Suppose that A is primitive. For each $e, e' \in E$ there exists a directed path that starts with edge e', ends with edge e and that does not have any 2-hop cycles. The spectral radius of A, denoted here by $\rho(A)$, is then strictly larger than 1. Let $(r_e : e \in E)$ and $(l_e : e \in E)$ be respectively a right and a left eigenvector of A corresponding to the eigenvalue $\rho(A)$, suitably normalized so that $r_e > 0$, $l_e > 0$ for $e \in E$ and $\sum_{e\in E} r_e l_e = 1$. Define the *weighted in-degree* $i(v)$ and the *weighted out-degree* $o(v)$ of each sensor $v \in V$ as

$$\begin{aligned} i(v) &= \sum_{v' \in N(v)} r_{(v',v)} \\ o(v) &= \sum_{v' : v \in N(v')} l_{(v,v')}. \end{aligned}$$

THEOREM 6.1 *If A is primitive then for $v \in V$*

$$\lim_{k\to\infty} \hat{\pi}_k^v(H_m) = 0, \tag{6.9}$$

for each hypothesis $m \in \{1, 2, \cdots, M\}$ such that

$$\prod_{v\in V} \phi_v(H_m)^{o(v)} < \max_{m'} \left\{ \prod_{v\in V} \phi_v(H_{m'})^{o(v)} \right\}.$$

Proof. Define the matrix $W = [w_{e,e'}]_{E\times E}$ by setting $w_{e,e'} = r_e l_{e'}$ for $e, e' \in E$. Let

$$\alpha(k) = \sum_{j=0}^{k-1} \rho(A)^j.$$

By equality (6.8)

$$\lim_{k\to\infty} \frac{x_k^h}{\alpha(k)} = \lim_{k\to\infty} \sum_{j=0}^{k-1} \left(\frac{A^j}{\rho(A)^j} \right) \frac{\rho(A)^j}{\alpha(k)} u^h = Wu^h,$$

where the last equality follows since $\rho(A) > 1$ and

$$\lim_{k\to\infty} \left\| \rho(A)^{-k} A^k - W \right\|_{\infty} = 0$$

due to [12, Theorem 8.5.1]. The estimate $\hat{\pi}_k^v(h)$ of sensor v at step k therefore satisfies

$$\begin{aligned}\hat{\pi}_k^v(h) &\propto \phi_v(h) \exp\left(\sum_{v' \in N(v)} x_k^h(v', v) \right) \\ &= \phi_v(h) \exp\left(\alpha(k) \left(i(v) \sum_{v' \in V} u^h(v') o(v') + \varepsilon(k) \right) \right),\end{aligned}$$

where $\varepsilon(k) \to 0$ as $k \to \infty$. The conclusion of the theorem now follows since

$$\phi_v(h) \exp\left(\alpha(k) i(v) \sum_{v' \in V} u^h(v') o(v') \right) = \phi_v(h) \left(\prod_{v' \in V} \phi_{v'}(h)^{o(v')} \right)^{\alpha(k) i(v)}$$

and $\lim_{k\to\infty} \alpha(k) = \infty$ owing to $\rho(A) > 1$. This completes the proof.

We now turn to distributed hypothesis testing and interpret Theorem 6.1 in terms of the posterior distribution π. Consider first a symmetric structure for the graph G so that $o(v) = o(v') = \mu > 0$ for all $v, v' \in V$. It can be verified, for example, that an $n \times n$ torus as in Figure 6.1 provides one such structure, and that the matrix A is primitive if n is odd. The definition (6.4) of node potentials $\phi_v : v \in V$ then leads to

$$\prod_{v \in V} \phi_v(H_m)^{o(v)} = \left(\pi_o(H_m) \prod_{v \in V} f_m^v(x_v) \right)^{\mu},$$

for $m = 1, 2, \cdots, M$, and equality (6.9) indicates that for each sensor v, $\lim_{k\to\infty} \hat{\pi}_k^v(H_m) = 0$ for all m such that H_m is not a MAP estimate with respect to π. In particular if the MAP estimate is unique, then the sensors unanimously identify it, in the sense that only the corresponding entry in the local beliefs will be non-negligible. This conclusion does not hold for general graphs whose vertices may have different weighted out-degrees. In fact observations made at sensors with larger weighted out-degrees have more influence on the collective opinion. In this case the final consensus reflects the right MAP estimate for certain values of observations $(x_v : v \in V)$, but identifies wrong choices for others, as the numerical examples of the following section illustrates.

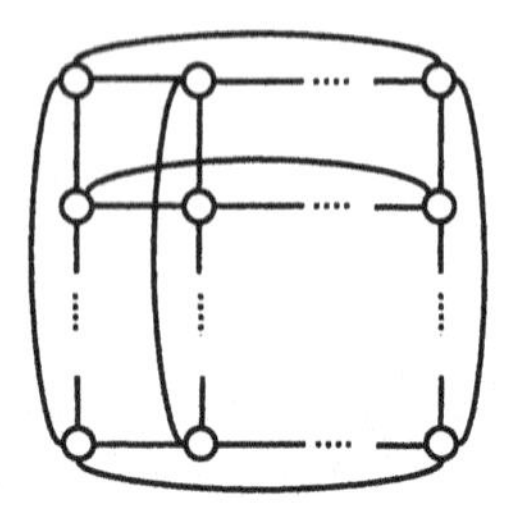

Figure 6.1. The torus topology in which $o(v) = o(v')$ for all nodes v, v'.

6. Numerical Examples

We start by illustrating the impact of unbalanced graphs on the final consensus of the network.

EXAMPLE 6.2 *Consider binary hypothesis testing in a 9-sensor network under the communication structure represented by the graph of Figure 6.2(a). Each edge in the graph represents two directed edges in opposite directions. The weighted out-degrees of the sensors with respect to this graph satisfy* $o(1) = o(2) = \cdots = o(8)$ *and* $o(0)/o(1) = 1.5091$. *Suppose that the observations* $(x_v : v \in V)$ *translate to node potentials* $\phi_0 = [q, 1-q]$, $\phi_1 = [p, 1-p]$ *and* $\phi_v = [0.5, 0.5]$ *for* $v = 2, 3, \cdots, 8$, *where* $p, q \in [0, 1]$. *Figure 6.2(b) illustrates the true MAP estimate and the final consensus due to belief propagation for different values of* p *and* q. *Note that the consensus is determined to a larger extent by the value of* q *rather than the value of* p. *Note also that the consensus reflects a flawed estimate if* (p, q) *lies in the area between the solid and dashed lines.*

We next illustrate the asymptotic behavior of the estimates $\hat{\pi}_k^v : v \in V$ in two topologies for which A is reducible. It should be noted here that the conclusions of the previous section continues to hold when A is reducible but non-primitive, though the analysis of this case is considerably more cumbersome. In the scope of the following three examples it is understood that for vertices $v, v' \in V$, $(v, v') \in E$ if and only if $(v', v) \in E$. Note that the undirected graph $\tilde{G} = (V, \tilde{E})$ is such that the unordered pair $[v, v'] \in \tilde{E}$ if and only if $(v, v') \in E$ and $(v', v) \in E$. The first example concerns the case when $\tilde{G}$ is a tree, and

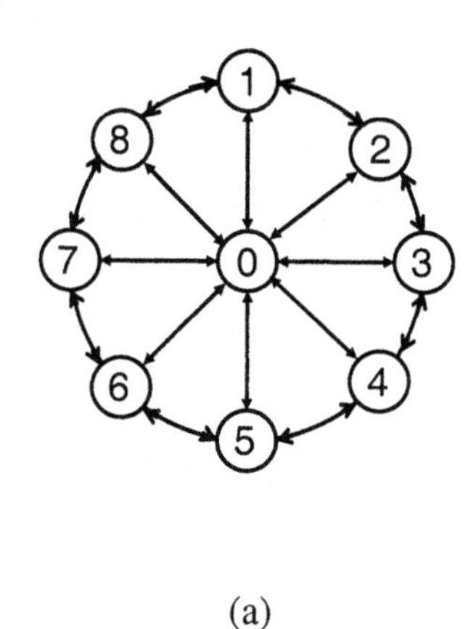

(a)

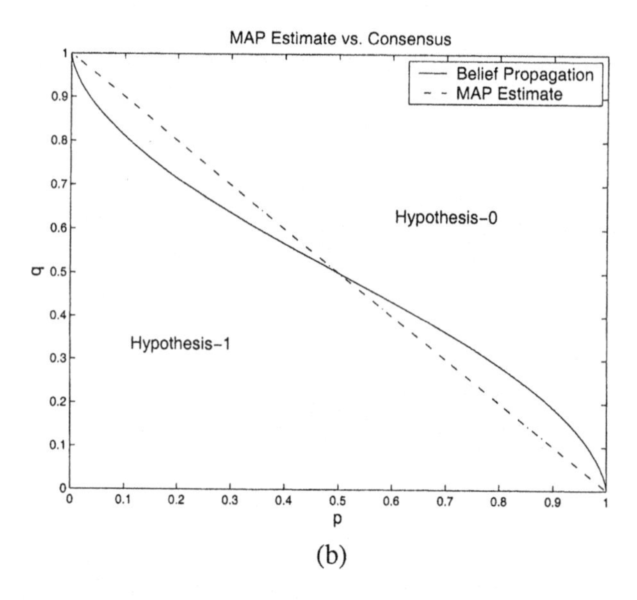

(b)

Figure 6.2. (a) The 9-node communication graph in which $o(1) = o(2) = \cdots = o(8)$ and $o(0)/o(1) = 1.5091$, (b) Decision regions for the MAP estimate and the final consensus of belief propagation, delineated respectively by the dashed line and the solid line. Observations $(x_v : v \in V)$ correspond to node potentials $\phi_0 = [q, 1-q]$, $\phi_1 = [p, 1-p]$ and $\phi_v = [0.5, 0.5]$ for $v = 2, 3, \cdots, 8$.

it is well-known that in this case belief propagation leads to the true posterior distributions for general Markov fields [26].

EXAMPLE 6.3 *(Trees) Suppose that $\tilde{G}$ is a tree, so that A is nilpotent since $A^j = 0$ for all integers j larger than the diameter of $\tilde{G}$. Equality (6.8) therefore indicates that the messages are guaranteed to converge within a number of steps no larger than the diameter of $\tilde{G}$. Note that for $\mathrm{e}, \mathrm{e}' \in E$*

$$\sum_{j=0}^{\infty} a^j_{\mathrm{e},\mathrm{e}'} = \begin{cases} 1 & \text{if there exists a simple directed path} \\ & \text{in } G \text{ with first edge } \mathrm{e}' \text{ and last edge } \mathrm{e} \\ 0 & \text{else,} \end{cases}$$

hence equality (6.8) leads to

$$\lim_{k\to\infty} x^h_k(e) = \sum_{v\in V} \mathbf{1}\{dist(v, s(e)) < dist(v, d(e))\} u^h(v),$$

for $\mathrm{e} \in E$, where $dist(v, v')$ represents the length of the unique path between vertices $v, v' \in V$. It now follows by equality (6.3) that the limit of the estimate $\hat{\pi}^v_k(h)$ at each sensor $v \in V$ is equal to the posterior distribution (6.1).

EXAMPLE 6.4 *(Rings) Suppose that $\tilde{G}$ is a simple cycle, so that for $\mathrm{e}, \mathrm{e}' \in E$ the sequence $(a^j_{\mathrm{e},\mathrm{e}'} : j = 0, 1, 2 \cdots)$ has period $|V|$. In particular $A^j = A^{j+|V|}$ and thus A is idempotent. Equality (6.8) then leads to*

$$\lim_{k\to\infty} \frac{x^h_k}{k} = \frac{1}{|V|} \sum_{j=0}^{|V|-1} A^j u^h.$$

It is not difficult to see that $\sum_{j=0}^{|V|-1} a^j_{\mathrm{e},\mathrm{e}'} = 1$ for all edges $\mathrm{e}, \mathrm{e}' \in E$ that have a common orientation (that is, clockwise or counter-clockwise) and that $\sum_{j=0}^{|V|-1} a^j_{\mathrm{e},\mathrm{e}'} = 0$ otherwise; in turn

$$\lim_{k\to\infty} \frac{x^h_k(e)}{k} = \frac{1}{|V|} \sum_{v\in V} u^h(v), \quad e \in E.$$

Therefore for large k the estimate $\hat{\pi}^v_k(h)$ of each sensor $v \in V$ at step k satisfies

$$\begin{aligned} \hat{\pi}^v_k(h) &\propto \phi_v(h) \exp\left(\sum_{v'\in N(v)} x^h_k(v', v) \right) \\ &\approx \phi_v(h) \exp\left(\frac{2k}{|V|} \sum_{v'\in V} u^h(v') \right) \\ &= \phi_v(h) \left(\prod_{v'\in V} \phi_{v'}(h) \right)^{\frac{2k}{|V|}}. \end{aligned}$$

Since

$$\prod_{v \in V} \phi_v(H_m) = \pi_o(H_m) \prod_{v \in V} f_m^v(x_v), \quad m = 1, 2, \cdots, M,$$

it follows that if H_m is not a MAP estimate with respect to π, then $\hat{\pi}_k^v(H_m) \to 0$ as $k \to \infty$. Note that if π leads to a unique MAP estimate m^* then the estimate distribution $\hat{\pi}_k^v$ of each sensor $v \in V$ converges so that

$$\lim_{k \to \infty} \hat{\pi}_k^v(H_m) = \mathbf{1}\{m = m^*\}, \quad m = 1, 2, \cdots, M.$$

In other words each sensor identifies the MAP estimate, although the limit of $\hat{\pi}_k^v$ is not necessarily the correct posterior distribution. MAP

The final example aims to shed light on the case when the MAP estimate is not unique.

EXAMPLE 6.5 *(4-ring) Consider a binary hypothesis testing problem involving 4 sensors arranged on a ring. Suppose that the observations $(x_v : v \in V)$ translate to node potentials $\phi_0 = [0.65, 0.35]$, $\phi_1 = [0.3, 0.7]$, $\phi_2 = [0.5, 0.5]$, $\phi_3 = [0.55, 0.45]$, so that H_1 is the unique MAP estimate in the centralized solution. The decentralized solution identifies the same MAP estimate since the estimate of the posterior distribution at each sensor converges to $(0, 1)$ as illustrated in Figure 6.3(a). If the node potentials are $\phi_0 = [0.7, 0.3]$, $\phi_1 = [0.4, 0.6]$, $\phi_2 = [0.6, 0.4]$, $\phi_3 = [0.3, 0.7]$, then the posterior distribution π assigns equal probabilities to both hypotheses. In this case the decentralized estimates $\hat{\pi}_k^v$ display oscillations around the correct probabilities as shown in Figure 6.3(b). Note that in this case also, only the correct MAP estimates display belief values that are asymptotically bounded away from zero.*

7. Conclusion

We have considered in-network data processing to classify an unknown event based on noisy sensor measurements. Belief propagation is applied as a message passing strategy to solve a distributed hypothesis testing problem for a pre-specified network connectivity. Reformulating the message evolution leads to a linear dynamical system model, which in turn allows characterizing conditions for reaching a consensus, and deriving conditions for when the consensus converges to the centralized MAP estimate. Main features of the proposed architecture are as follows:

(i) Ad-hoc communication infrastructure. The sum-product algorithm converges here to the correct MAP estimate under a rich variety of communication topologies. The simplest such topology is a spanning tree, and any balanced topology delivers the desired estimate. Unbalanced topologies yield correct

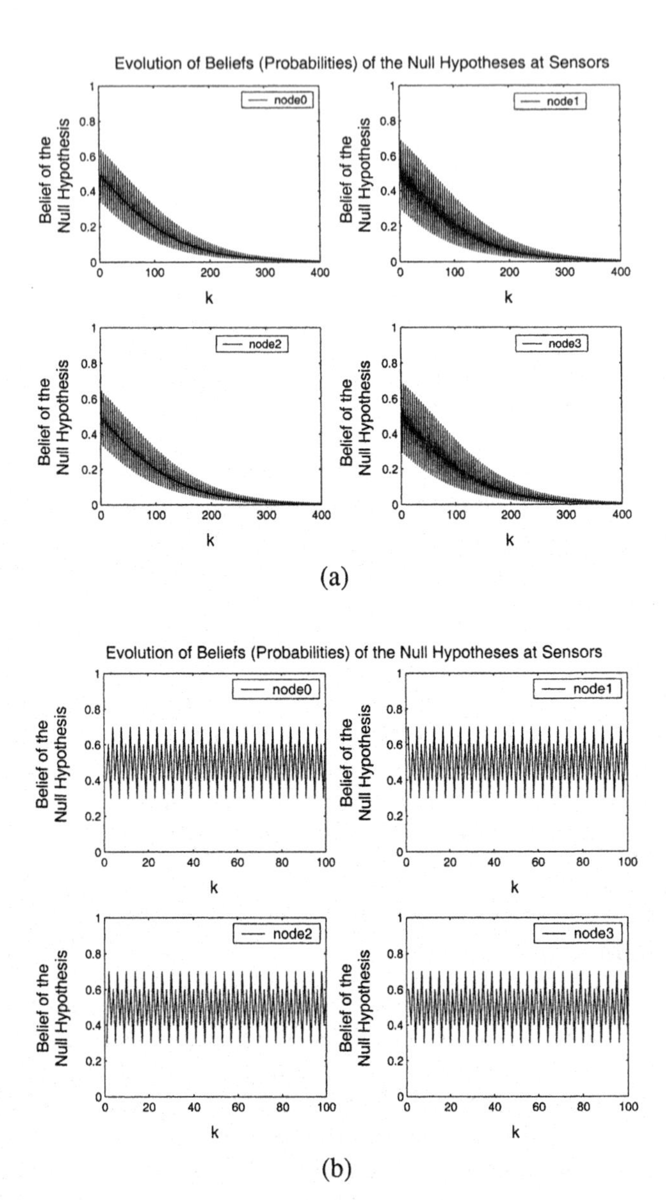

Figure 6.3. Estimates of the posterior distribution in a ring of 4 sensors. The node potentials, in clockwise order, are (a) $\phi_0 = [0.65, 0.35]$, $\phi_1 = [0.3, 0.7]$, $\phi_2 = [0.5, 0.5]$, $\phi_3 = [0.55, 0.45]$, and (b) $\phi_0 = [0.7, 0.3]$, $\phi_1 = [0.4, 0.6]$, $\phi_2 = [0.6, 0.4]$, $\phi_3 = [0.3, 0.7]$.

estimates for wide range of observation values, and exact quantification of this entails assessing how unbalanced the topology is in terms of the weighted out-degrees of sensors.

(ii) Robustness to sensor failures. Typical applications of sensor networks involve inexpensive and unreliable sensors. The performance of the proposed architecture degrades gracefully with sensor failures, which amount to topological variability. This feature of the architecture stands in contrast to tree-based in-network data aggregation schemes [36–40] that are vulnerable to single points of failure.

(iii) Robustness versus efficiency trade-off. The robustness of the proposed scheme entails increased messaging complexity, which is a crucial factor that determines energy efficiency of power-limited sensors. If, however, dynamic situations are considered where the sensing task is tracking rather than one-shot classification, then the amortized messaging complexity is a more appropriate measure of energy efficiency. This aspect of the present scheme is currently under study.

Operational advantages of the considered architecture are:

(i) Local information. A key feature of the present architecture is that individual sensor operation is based on local information. Namely, each message is determined locally by the observation at the sensor and the prior messages received from neighboring sensors, and the message forwarding does not require global knowledge of sensor models. This stands in sharp contrast with previously proposed decentralized estimation schemes where such global knowledge is required at the fusion center.

(ii) Network-wide consensus. The network eventually settles to a state of consensus in which all sensors identify the same MAP estimates. Final estimates can therefore be collected by probing an arbitrary sensor, leading to further operational flexibility in interfacing with the control plane.

(iii) Scalability due to event-driven operation. The operation of each sensor is event-driven, namely it is driven by message exchanges among neighbors. Hence network-wide synchronization is not required and implementation scales naturally to large sensor networks.

References

[1] P. Bergamo, A. Giovanardi, A. Travasoni, D.Maniezzo, "Distributed power control for energy efficient routing in adhoc networks," Wireless Networks, vol. 10, pp. 29–42, 2004.

[2] T. Berger, Z. Zhang, and H. Vishwanathan, "The CEO problem," *IEEE Transactions on Information Theory*, May 1996.

[3] V. Borkar, P. Varaiya, "Asymptotic agreement in distributed estimation," IEEE Transactions on Automatic Control, 27, 1982,

[4] M.S. Branicky, V.S. Borkar, S.K. Mitter,"A unified framework for hybrid control: Model and optimal control theory," IEEE Trans. Automatic Control, 43(1):31-45, January 1998.

[5] J. F. Chamberland and V. V. Veeravalli, "Decentralized detection in sensor networks," *IEEE Transactions on Signal Processing*, 2003.

[6] "Collaborative information processing," *IEEE Signal Processing Magazine*, 2002.

[7] R.L. Cruz and A.V.Santhanam, "Optimal routing link scheduling and power control in multi-hop wireless networks," Infocom 2003.

[8] D. Estrin, D. Culler, K. Pister, and G. Sukhatme, "Connecting the physical world with pervasive networks," *IEEE Pervasive Computing*, vol. 1, pp. 59–69, January-March 2002.

[9] M. Gastpar, P. L. Dragotti, M. Vetterli, "The distributed, partial and conditional Karhunen-Loeve transforms," In Proc 2003 IEEE Data Compression Conference, March 2003

[10] Y. C. Ho,"Team Decision Theory and Information Structures," Proc. IEEE, 68, 1980

[11] S. E.-A. Hollar, "COTS dust," Master's thesis, University of California, Berkeley, 2000.

[12] R. A. Horn and C. R. Johnson, *Matrix Analysis*, Cambridge University Press, 1985.

[13] G. T. Huang, "Casting the wireless sensor net," *MIT Technology Review*, July-August 2003.

[14] P. Ishwar, R. Puri, S. S. Pradhan and K. Ramchandran, "On rate-constrained estimation in unreliable sensor networks," Proc. 2nd International workshop on Information processing in sensor networks (IPSN), April 2003

[15] J. M. Kahn, R. H. Katz, and K. S. J. Pister, "Next century challenges: mobile networking for 'smart dust'," in *Proceedings of the fifth annual ACM/IEEE international conference on Mobile computing and networking*, (Seattle, WA, United States), ACM, 1999.

[16] IETF Mobile Ad-hoc Networks Working Group. http://www.ietf.org/html.charters/manet-charter.html.

[17] R. J. McEliece, D. J. C. MacKay, and J.-F. Cheng, "Turbo decoding as an instance of Pearl's "belief propagation" algorithm," Journal on Selected Areas in Communications, vol. 16, no. 2, pp. 140–151, Feb. 1998.

[18] S.K. Mitter, N.J. Newton, "Information Flow and Entropy Production in the Kalman-Bucy Filter," LIDS publication no. P-2577, submitted to J. of Stat. Phys., February 2004

[19] P. Mohapatra, J. Li, and C. Gui, "Qos in mobile ad hoc networks: Special issue on qos in next-generation wireless multimedia communications systems," *IEEE Communications Magazine*, pp. 11–21, June 2003.

[20] M.J. Neely, E. Modiano, and R.E. Rohrs, "Dynamic power allocation and routing for time varying wireless networks," Infocom 2003.

[21] R. Nowak and U. Mitra, "Boundary estimation in sensor networks," in *2nd International Workshop on Information Processing in Sensor Networks*, (Palo Alto, CA), April 2003.

[22] R. Nowak, "Distributed em algorithms for density estimation and clustering in sensor networks." IEEE Transactions on Signal Processing, Special Issue on signal Processing in Networking, 2003.

[23] Y. Oohama, "The rate distortion function for the quadratic gaussian ceo problem," *IEEE Transactions on Information Theory*, vol. 44, pp. 1057–1070, May 1998.

[24] N. Patwari and A. Hero, "Reducing transmissions from wireless sensors in distributed detection networks using hierarchical censoring," in *ICASSP*, IEEE, 2003.

[25] C. Perkins, "Ad-hoc Networking," Addison-Wesley, 2000.

[26] J. Pearl, "Probabilistic reasoning in intelligent systems: networks of plausible inference," Morgan-Kauffman, 1988.

[27] S. Pradhan, J. Kusuma, and K. Ramchandran, "Distributed compression in a dense microsensor network," IEEE Signal Processing Magazine, 2002.

[28] B. Radunovic and J. Le Boudec, "Joint scheduling, power control, and routing in symmetric one-dimensional multi-hop wireless networks," WiOpt, INRIA, Nice, France, March 2003.

[29] Suresh Singh, Mike Woo, and C. S. Raghavendra, "Power-Aware Routing in Mobile Ad Hoc Networks," in Mobile Computing and Networking, pp. 181-190, 1998.

[30] J. N. Tsitsiklis and M. Athans, "Convergence and asymtotic agreement in distributed decision problems," *IEEE Trans. Automatic Control*, 1984.

[31] J. N. Tsitsiklis, "Decentralized detection," *in Advances in Statistical Signal Processing, H. V. Poor and J. B. Thomas Eds*, vol. 2.

[32] N. Vaidya, "Tutorial on mobile ad-hoc networks." Available at: Available at: `http://www.crhc.uiuc.edu/nhv/presentations.html`.

[33] H. L. Van Tress, *Detection Estimation and Modulation Theory*, John Wiley and Sons, 1968.

[34] P. K. Varshney, *Distributed Detection and Data Fusion*. Springer, 1997.

[35] H. Xiaoyan, X. Kaixin, and M. Gerla, "Scalable routing protocols for mobile ad hoc networks," *IEEE Network Magazine*, pp. 11–21, July-August 2002.

[36] Y. Yao and J. Gehrke, "The Cougar approach to in-network query processing in sensor networks," Sigmod Record, Volume 31, Number 3, September 2002.

[37] J. Beaver, M.A. Sharaf, A. Labrinidis, and P.K. Chrysanthis, "Power-aware in-network query processing for sensor data," Proceedings of the 2nd Hellenic Data Management Symposium, September 2003.

[38] J. Liu, J. Reich, and F. Zhao, "Collaborative in-network processing for target tracking," Journal on Applied Signal Processing, vol. 23, no. 4, pp. 378-391, March 2003.

[39] R. Kumar, V. Tsiatsis, and M.B. Srivastava, "Computation hierarchy for in-network processing," Proceedings of the Second International Workshop on Wireless Sensor Networks and Applications, September 2003.

[40] J. Considine, F. Li, G. Kollios and J.W. Byers, "Approximate aggregation techniques for sensor databases," 20th IEEE International Conference on Data Engineering (ICDE '04), 2004.

Chapter 7

MAXIMIZING AGGREGATE THROUGHPUT IN 802.11 MESH NETWORKS WITH PHYSICAL CARRIER SENSING AND TWO-RADIO MULTI-CHANNEL CLUSTERING*

Jing Zhu, Sumit Roy
Department of Electrical Engineering
University of Washington, Seattle, WA 98195
{zhuj,roy}@ee.washington.edu

Xingang Guo, and W. Steven Conner
Communications Technology Lab, Intel Corporation, Hillsboro, OR 97124
{xingang.guo,w.steven.conner}@intel.com

Abstract Spatial reuse in a mesh network allows multiple communications to proceed simultaneously, hence proportionally improving the overall network throughput. To maximize spatial reuse, the MAC protocol must enable simultaneous co-channel transmitters to maintain a separation distance that is sufficient to avoid interference. Within that distance, a set of orthogonal channels is employed by different links. This paper demonstrates that physical carrier sensing enhanced with a tunable sensing threshold is effective at avoiding co-channel interference in 802.11 mesh (static + multi-hop) networks. Moreover, for multi-channel mesh networks, an architecture for channel clustering based on two-radio nodes is proposed. Distributed clustering is achieved using the Highest-Connectivity Cluster (HCC) algorithm. All inter-cluster communications are performed on a common channel using the default radio, while intra-cluster communications use the secondary radio with channel selection based on a new Minimum Interference Channel Selection (MIX) algorithm that minimizes the co-channel interference (CCI). Backward compatibility is guaranteed by allowing legacy single-channel devices to connect to the new two-radio devices through the common default radio. Simulation results for large-scale 802.11b and 802.11a networks demonstrate the significant improvement in one-hop aggregate throughput. Specifically,

*The work of Jing Zhu and Sumit Roy was supported in part by an Intel Research Council Grant.

the new two-radio multi-channel mesh solution increases the aggregate throughput by more than twice w.r.t. the traditional single-radio single-channel mesh.

Keywords: Media Access Control, clustering, 802.11.

1. Introduction

The past few years have witnessed the rapid proliferation of wireless LANs in various environments: home, enterprise and hotspot. The need for higher data rates and improved coverage has led to multi-cell networks (particularly for business and hotspot scenarios but also for clusters of homes/apartments) where each cell is served by its own access point (AP). Currently, all APs are directly connected (typically via Ethernet) to an Internet gateway. Therefore, the cost and time of deploying a large scale WLAN network dramatically increases as the network expands. A possible solution to this problem is connecting APs wirelessly to form a (static) multi-hop .11 (*mesh*) network (see Fig.7.1). The high interest in such an approach is indicated not only by the newly formed Mesh Task Group within IEEE 802.11 but also mesh solutions offered by several companies [2–4] to list a few. Such a future *wireless AP-AP mesh network* requires both protocol and architectural extensions to current .11 networks for which there does not exist any standardized inter-AP connectivity protocol. Furthermore, since the wireless channel is a broadcast (shared) medium with bandwidth limitations, the aggregate throughput of such a wireless inter-AP mesh is governed by the following key network parameters:

K: number of concurrent active links per channel (degree of co-channel spatial reuse);

W: max. data rate per channel;

N: number of orthogonal channels within a reuse distance

This paper looks at approaches to maximize K as well as utilizing N to maximize the aggregate throughput of a wireless mesh network.

In [7], spatial reuse was demonstrated to depend on various characteristics of the network, including the type of radio, network topology, channel quality requirements and signal propagation environment. For a given network configuration, there exists a minimum separation distance such that when simultaneous transmitters are separated by that distance, the maximum number of simultaneous transmissions can be accommodated, allowing maximum network throughput to be achieved. However, achieving maximum spatial reuse would require an ideal MAC protocol that schedules communication to maintain the optimal transmitter separation distance (to minimize interference) in a fully distributed manner.

Nodes in a IEEE 802.11 WLAN network seeking channel access use carrier sensing to determine if the medium is available before transmitting to avoid packet collision [1]. Two types of carrier sensing are supported by the 802.11 MAC: mandatory physical carrier sensing that monitors the RF energy level in the air and optional virtual carrier sensing that uses the Request-to-Send/Clear-to-Send (RTS/CTS) handshake to ensure that the air medium at the receiver is reserved prior to data packet transmission. Virtual carrier sensing was designed to avoid the well-known hidden terminal problem [11], where it is assumed that physical carrier sensing at a transmitter is not sufficient to avoid interference at a receiver. However, it has been shown that virtual carrier sensing via RTS/CTS in fact suffers from fundamental limitations in avoiding interference from hidden terminals [12]. In 802.11, this can be attributed to lack of proper design of the physical carrier sensing mechanism. In this paper we demonstrate that, when properly tuned, physical carrier sensing is effective at avoiding interference in a multi-hop wireless mesh network *without the use of virtual carrier sensing*.

Physical carrier sensing allows a station to assess the channel condition before transmitting to make sure that no interference can occur. A node samples the energy on the channel and initiates channel access only if the reading is below the carrier sensing threshold, indicating that any ongoing communication only produces tolerable interfere with the impending transmission. According to RF pathloss models, the long-term average received energy at a node decays with distance from a transmitter. Hence the carrier sensing threshold effectively determines the minimum allowed distance between simultaneous transmitters. Since the optimal distance depends on various network properties, the carrier sensing threshold should be tuned to current network conditions. However, many of today's 802.11 MAC implementations use a static threshold, or do not allow the threshold to be independently tunable [16]. As a result, physical carrier sensing often leads transmitters to be either too conservative or too aggressive when using the wireless channel.

In this work, we assume a tunable carrier sensing threshold and illustrate how to derive the appropriate carrier sensing threshold from relevant network characteristics via analysis. Furthermore, we propose an estimation-based adaptive physical carrier sensing scheme to automatically tune the threshold to a near-optimal value. We present OPNET simulation results for two regular network topologies (chain and grid) to validate the theoretical optimal PCS threshold. Our results further show that by tuning the physical carrier sensing threshold, without requiring virtual carrier sensing, the overall network throughput can be improved significantly compared to that of the legacy 802.11 MAC. The increased throughput can approach approximately 90% of the theoretical upper-bound predicted by spatial reuse models in a large chain. Simulation results also demonstrate the effectiveness of the estimation-based adaptive physical carriers sensing scheme in networks with dynamic topology and heterogeneous links.

We note that performance improvement of 802.11 networks based on enhancement to various aspects of the 802.11 MAC protocol has been the subject of recent work [10] [14]. For any given environment, optimizing network performance must be a careful combination of approaches addressing multiple aspects of network performance (e.g. throughput, fairness, etc.) which is beyond the scope of this work. We focus here specifically on *leveraging the spatial reuse* of mesh networks to enhance the throughput through physical carrier sensing, which is an essential requirement for achieving optimal aggregate throughput in a dense wireless network.

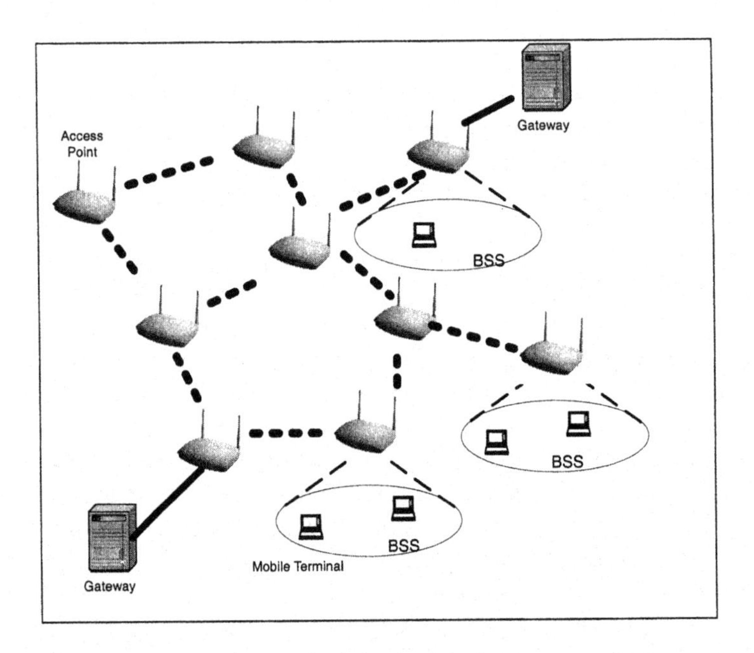

Figure 7.1. Wireless AP-to-AP mesh Networks

Communication between nodes on the AP-mesh can share a single channel or use multiple narrower-band channels. This can be implemented readily using a single-radio network (all nodes have only one radio interface) that suggests the need for a single *wideband* shared channel for the entire AP-mesh to support many simultaneous transmissions. However, this approach has not been adopted by industry standards as yet and lacks hardware implementations [1]. Using multiple narrower-band channels entails potentially complicated channel assignment schemes to inter-AP links, but has existing hardware support (N=8 in 802.11 a, and N=3 in 802.11 b). Thus our proposed solution is based on using a $20MHz$ channel for all inter-AP communications while noting that this is likely to be a throughput bottleneck in situations where the inter-AP (routed) traffic dominates.

Two-radio multi-channel approaches, where each node is equipped with two similar PHY/MAC radio interfaces for AP-AP communication, can effectively exploit the available multiple orthogonal channels [2] which is infeasible with a one-radio AP Mesh with multiple narrowband channels. In a two-radio implementation, each node in the AP-mesh is equipped with two WLAN cards that are used for intra-cell and inter-cell communications, respectively. This will always provide higher throughput than a single-channel approach, since intra-AP traffic is now separated from inter-AP traffic made possible by two-radio nodes. Such a multi-radio architecture has been proposed in [9] where the channel used for any AP-AP link from among the available set is selected by sending probes to estimate the link round trip time (RTT) for the available channels and choosing the one with minimum RTT. Updating the channel allocation is performed periodically every few seconds. While the RTT is a useful indicator of channel load, it is a less-than-adequate metric for estimating interference due to simultaneous transmissions in a wireless scenario. Thus [9] protocol operates more as a load-balancing scheme which improves but does not optimize aggregate network throughput.

Continuous monitoring of channel quality on all channels is infeasible with a single radio; two radios per node considerably simplifies this because this task can be performed by one radio while the other is transmitting data on the currently assigned channel. Suggestions for using one radio purely as a dedicated control channel and the other for data on all other channels have appeared for two-radio architectures [26] [25]; but these lead to low channel utilization due to control channel becoming a bottleneck, and offers no backward compatibility. Thus in our implementation, both radios are used to support data transmission. Nonetheless, irrespective of the specifics of how the two radios are used, this architecture allows the possibility of a *fully distributed* MAC implementation that is desirable for network robustness. For example, to eliminate the control channel bottleneck, we propose a new *semi-distributed* AP-clustering approach. A distributed Highest-Connectivity Cluster (HCC) algorithm [22] is employed to divide the network into AP clusters that are distinguished by the channel used for intra-cluster communication. Inter-cluster communication is performed using the (*default* and intra-cluster via the *secondary*) radio, respectively.

A common channel is used for all inter-cluster communications, and different channels are selected for intra-cluster communications by using a new Minimum Interference Channel Selection (MIX) algorithm. Control or management traffic uses only the default radio; while the secondary radio is only for data transmissions. Note that backward compatibility is achieved since this architecture allows a legacy single-radio AP to connect to the new two-radio APs through the (common) default radio.

Unlike most of other cluster-based networks (e.g. Bluetooth, UWB) that usually employ a cluster head as a controller running a centralized MAC, the architecture here uses clustering to only assign a channel in a distributed manner for the MAC; the base 802.11 MAC mechanisms are unchanged. Similar to [9], our protocol uses a virtual MAC address in place of the multiple physical MAC addresses used by two radios so that the higher (routing) layer sees only a single wireless network interface. Routing between the nodes is based on ad-hoc routing approaches similar to that in the traditional single-channel, single-radio mesh.

The rest of this paper is organized as follows: Section 2 presents a basic communication model for interference analysis and exposes the limitations of carrier sensing as currently implemented in 802.11 DCF. Section 3 introduces our suggestions for enhanced physical carrier sensing based on a tunable physical sensing threshold, and demonstrates the resulting throughput improvements for some regular mesh networks topologies. Section 4 describes our novel two-radio multi-channel clustering architecture. Section 5 presents OPNET simulation results showing significant performance enhancements obtained from tuned physical carrier sensing along with new clustered two-radio, multi-channel AP-mesh. Section 6 discusses our results in the context of related work, and Section 7 concludes the paper.

2. Managing Interference with Physical Carrier Sensing (PCS)

In CSMA/CA based wireless networks such as IEEE 802.11 networks, a transmitter relies on carrier sensing to determine if the medium is 'available', i.e., has acceptable level of interference from ongoing transmissions. A transmission is initiated only if the energy level is below the PCS threshold. This section uses common radio propagation models to determine the effectiveness of carrier sensing and points out several shortcomings of the carrier sensing technique employed in 802.11 MAC protocol.

2.1 Communication Model

Path loss models are commonly used to describe the average received power at a receiver over a wireless medium [7] [19] as a function of the T-R (transmitter-receiver) radial separation, d, i.e.,

$$P_{rx}(d) = \bar{P}_{rx}(\frac{\bar{d}}{d})^{\gamma} \quad (7.1)$$

where γ is the path loss exponent that characterizes the rate of signal degradation with distance in the particular network environment. $P_{rx}(d)$ denotes the signal power at distance d from the transmitter and $\bar{P}_{rx}$ is the signal power measured at a reference distance $\bar{d}$ (usually 1 meter).

The aggregate energy at any receive node consists of desired signal, interference (from unwanted transmitter(s)) and background noise. A 802.11 node can receive a packet with high probability of success in additive noise only if the received signal strength is greater than a threshold (denoted by P_R, i.e. reception sensitivity); and equivalent condition in the presence of interference is that the received Signal-Interference and Noise-Ratio (SINR) exceeds a threshold denoted by S_0; i.e.,

$$\begin{cases} P_{rx}(d) \geq P_R \\ \frac{P_{rx}(d)}{P_N + \sum_i P_{rx}(d_i)} \geq S_0 \end{cases}, \tag{7.2}$$

where P_N is the background noise power, and $P_{rx}(d_i)$ denotes the power of interference source i at distance d_i from the receiver. 802.11 networks support multiple data rates, and a higher data rate requires a higher S_0.

2.2 Terminologies

Eq.7.2 provides constraints on the receive power as well as the SINR at detector input for successful detection. In the absence of any interference, the receive sensitivity P_R is set to satisfy $P_R/P_N > S_0$; this determines the *transmission range* or the maximum distance for successful reception in additive noise only. It is clear that the actual SINR perceived via PCS at a receive node will vary due to the presence of interference from ongoing transmissions.

Fig.7.2 shows a typical mesh network with a reference transmission from a node TX to a node RX in the presence of four other nodes (A, B, C, and E). The same transmission power is used by every node in the network. We define the following:

D: TX-RX separation distance, defines $P_D = P_{rx}(D)$.

R: Transmission range, given by

$$R = \bar{d}\left(\frac{\bar{P}_{rx}}{\max(P_R, S_0 P_N)}\right)^{\frac{1}{\gamma}} = \bar{d}\left(\frac{\bar{P}_{rx}}{P_R}\right)^{\frac{1}{\gamma}}, \tag{7.3}$$

I: Interference range – implies a single transmitter within that range of the receiver will disrupt reception of the desired transmitter, given by

$$I = D\left(\frac{1}{\frac{1}{S_0} - (\frac{D}{\bar{d}})^{\gamma}\frac{P_N}{\bar{P}_{rx}}}\right)^{1/\gamma}. \tag{7.4}$$

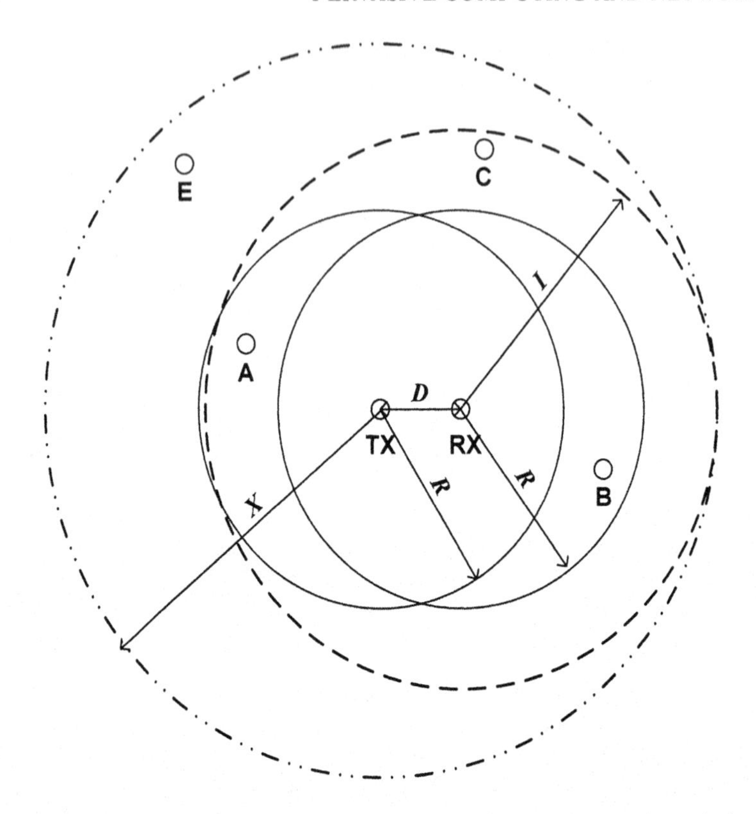

Figure 7.2. Illustration of relative transmission and interference distances in a wireless mesh network.

With negligible background noise, Eq.7.4 turns to

$$I \approx S_0^{1/\gamma} D. \tag{7.5}$$

X: Physical carrier sensing range – a node will be able to detect an existing transmitter within that range via physical carrier sensing, given by

$$X = \bar{d}(\frac{\bar{P}_{rx}}{P_C})^{\frac{1}{\gamma}}, \tag{7.6}$$

where P_C denotes the physical carrier sensing (PCS) threshold.

Table 7.1 briefly summarizes the common symbols used throughout this paper to describe carrier sensing.

Table 7.1. Common symbols for describing physical carrier sensing.

Symbol	Description
P_R	Received Power Threshold
P_C	Physical Carrier Sensing Threshold
P_D	Received Power at distance D
P_N	Background Noise Power
P_I	Interference Power
S_0	SNIR Threshold
γ	Path Loss Exponent
X	Physical Carrier Sensing Range
R	Transmission Range
I	Interference Range
D	Transmission Distance
p_{cs_t}	P_C/P_D (normalized CS threshold)
k	Spatial reuse factor
W	Link capacity

2.3 Limitations of Carrier Sensing in 802.11 MAC Protocols

In today's 802.11 networks, the PCS scheme is typically configured with a fixed threshold, which is often set very low such that even a remote communication would force a station to withhold its transmission. Clearly, dynamic tuning of the PCS threshold according to current network conditions allows for optimum exploitation of spatial capacity.

In addition to PCS, 802.11 MAC also allows for virtual carrier sensing (VCS) scheme [1] for interference avoidance. This is accomplished by a initial Request-to-Send (RTS)/ Clear-to-Send (CTS) handshake prior to data transmission. With VCS, each station maintains a NAV (Network Allocation Vector) that indicates the period(s) during which the shared medium is reserved by other stations, hence it knows when NOT to transmit. When contending for the medium, a station broadcasts its intended transmission period in the RTS or CTS; each station that receives the broadcast updates its NAV. Hence, VCS requires participating stations to be able to receive and decode the RTS/CTS broadcast frames from any other stations in the network with which they may potentially interfere. Unfortunately, this requirement cannot be guaranteed in many scenarios [12] as exemplified in Fig. 7.2. The VCS scheme appropriately prevents nodes A and B from initiating an interfering transmission, as they are in the transmission range of TX and RX. But node C is too far away from both TX and RX to reliably receive and decode the RTS or CTS packets, yet it is still a potential hidden node that could interfere with the packet reception at RX.

3. Enhancing Physical Carrier Sensing

3.1 Tuning Physical Carrier Sensing (PCS) to Avoid Interference

Physical carrier sensing allows a station to assess the channel conditions before transmitting to avoid interference that will lead to packet collisions. A station samples the energy level at the air interface and starts a packet transmission only if the energy level is below a threshold P_C, called the PCS threshold.

In a multi-hop wireless network, only a subset (of the overall nodes) belonging to a cluster share a common transmission medium on a contention basis. The fundamental factor that determines whether a packet can be successfully received is the SINR at the receiver. If the signal that a device is attempting to receive is sufficiently stronger than the background noise and interference, successful packet reception can occur even in the presence of interference. Thus, the goal of PCS is to prevent those simultaneous transmissions that will lead to packet collisions, while permitting other simultaneous transmissions that do not violate receive SNIR requirements and thus maximize spatial reuse.

Fig. 7.3 illustrates a simple example of how the choice of PCS threshold can impact wireless network performance. If the threshold is too high, the CSMA is needlessly conservative. While node C is transmitting, both nodes A and B will backoff as in Fig. 7.3(a) , even though node A may be able to simultaneously transmit without causing much interference at C's receiver to disrupt successful communication. On the other hand, as shown in Fig. 7.3 (b), if the threshold is too high so as to allow both nodes A and B to transmit simultaneously with C, excessive interference will be generated resulting in packet collisions. If the PCS threshold is appropriately configured, as shown in Fig. 7.3 (c), nodes A and C will be permitted to successfully transmit simultaneously while node B will be forced to back off to prevent packet collisions. When the PCS threshold is optimized, maximal spatial reuse can be achieved without permitting packet collisions.

When properly tuned, PCS is more robust than VCS, because it does not require control packets to be received and correctly decoded. It is also more flexible, since the PCS sensing range can be easily adjusted by tuning the PCS threshold. In Fig.7.2, all potentially interfering nodes, including node C, can be eliminated by enlarging the PCS sensing range to cover the entire potential interference area, i.e.

$$X \geq D + I. \tag{7.7}$$

Combining Eq.7.7 with Eq.7.5, we obtain

$$X \geq D(1 + S_0^{1/\gamma}), \tag{7.8}$$

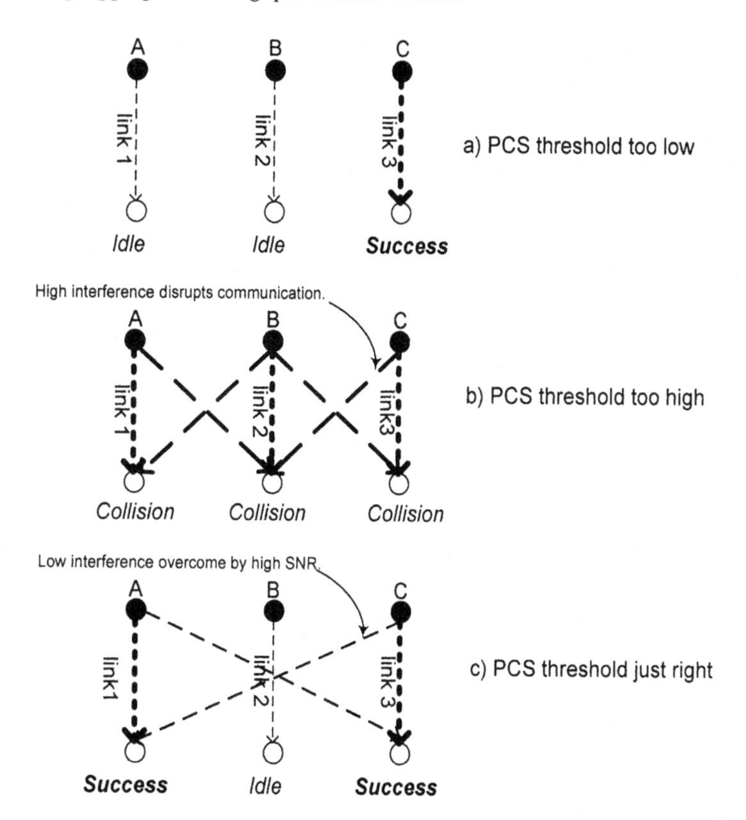

Figure 7.3. Physical Carrier Sensing (PCS) and Spatial Reuse.

that leads to

$$p_{cs\_t} \leq \frac{1}{(1 + S_0^{1/\gamma})^\gamma}. \tag{7.9}$$

A potential downside of this approach is the *exposed terminal problem* [10]. For example, in Fig. 7.2 even though a transmission by node E will not disrupt RX, E will defer its transmission because it lies within the sensing range. Having too many exposed terminals can potentially reduce the overall network throughput. However, by tuning the physical carrier sensing threshold, we will demonstrate a good tradeoff between solving the hidden terminal problem and exacerbating the exposed terminal problem, thereby obtaining high aggregate throughput.

3.2 Estimating Optimal PCS Threshold to Maximize Spatial Reuse

As already motivated by the earlier example, choosing the optimal PCS threshold can maximize spatial reuse leading to increased aggregate network one-hop throughput. In order to establish some preliminary guidelines for the choice of an optimal PCS threshold, we assume a homogeneous network with identical interference environment at each node [3]. The optimal spatial reuse is achieved when the number of simultaneous successful transmissions reaches the maximum. For successful reception at a receive node, the net interference and noise cannot exceed the tolerable level according to Eq.7.2,

$$P_I + P_N \leq P_D/S_0, \tag{7.10}$$

With the assumption that the transmitter and the receiver perceives the same interference and noise level, the optimal PCS threshold should satisfy

$$P_C \leq P_D/S_0, \tag{7.11}$$

for successful simultaneous transmissions. Hence, P_D/S_0 is the optimal PCS threshold for maximum spatial reuse; a higher P_D/S_0 implies more simultaneous transmissions and greater reuse. The corresponding optimal p_{cs_t} denoted as β, is then

$$\beta = \frac{1}{S_0} \tag{7.12}$$

independent of path loss exponent γ.

Recall Eq.7.9 that provides a necessary condition for completely eliminating the interference from hidden nodes. The ratio ρ of the exposed terminal area to the whole PCS sensing area is given by

$$\rho = \frac{\pi X^2 - \pi I^2}{\pi X^2} \approx \frac{D^2(1+S_0^{1/\gamma})^2 - D^2 S_0^{2/\gamma}}{D^2(1+S_0^{1/\gamma})^2} = 1 - \left(\frac{S_0^{1/\gamma}}{1+S_0^{1/\gamma}}\right)^2. \tag{7.13}$$

When $S_0^{1/\gamma}$ is small, ρ is not negligible; but for $S_0^{1/\gamma} >> 1$ [4], we have $\rho \approx 0$ so that the exposed terminal problem can be ignored, and Eq.7.9 reduces to $p_{cs\_t} \leq \beta$.

3.3 Analysis Model for Aggregate Throughput Limits

In [7], spatial reuse for a homogeneous ad-hoc environment was investigated where every transmitter uses the same transmission power and data rate, and communicates to an immediate neighbor at the constant T-R distance d. The spatial reuse can be characterized by the distance between neighboring simultaneous transmitters (T-T separation). The optimal spatial reuse (min. T-T separation) for two regular network topologies: the 1-D chain network and the 2-D grid network were derived. Let k denote the T-T distance (also called spatial reuse factor) measured in number of hops (hop distance d equals inter-node separation), then k must satisfy

$$\begin{cases} k \geq \left[2\left(1+\frac{1}{\gamma-1}\right)S_0\right]^{\frac{1}{\gamma}}, & \text{Chain network} \\ k \geq \left[6\left(1+\frac{1}{\gamma-2}\right)S_0\right]^{\frac{1}{\gamma}}, & \text{2-D grid} \end{cases} \tag{7.14}$$

We assume that a suitable MAC protocol schedules simultaneous communication only for transmitters that are k hops away; the network then reaches its aggregate throughput limit. In a chain network of N nodes, a packet must be relayed by each of the $N-2$ intermediate nodes in order to be routed from one end to the other. Since at most N/k simultaneous transmitters can be supported in the chain, the end-to-end throughput C_{e2e} is approximated by

$$C_{e2e} \approx \frac{W}{N} \times \frac{N}{k} = \frac{W}{k} \tag{7.15}$$

where W denotes the link capacity.

4. A Multi-Channel Two-Radio Architecture with Clustering

A multi-channel architecture with clustering was previously studied in [23], which only considered a centralized TDMA MAC and one radio. Here, we propose to integrate two 802.11 radios (*default* and *secondary*) per node: the default radio is used for inter-cluster communications; while the secondary radio is for intra-cluster communications. Unlike most existing multi-channel approaches, the new clustered multi-channel two-radio (CMT) architecture not only eliminates the need for switching channels on a packet-by-packet basis, but is also fully compatible with legacy devices. Fig.7.4 shows an example of a mesh network using the CMT architecture with three orthogonal channels, where each circle represents an independent cluster.

Fig. 7.5 shows protocol stack in a two-radio device. We highlight the two new modules - MAC Extension and Secondary MAC/PHY - needed to enable the two-radio functionality. Algorithms in the new architecture are implemented in the MAC Extension. The secondary MAC/PHY has no administrative functionality, such as association, authentication etc. and can transmit only data traffic.

Clustering is accomplished by using the Highest-Connectivity Cluster (HCC) algorithm first proposed in [22], based on the following rules:

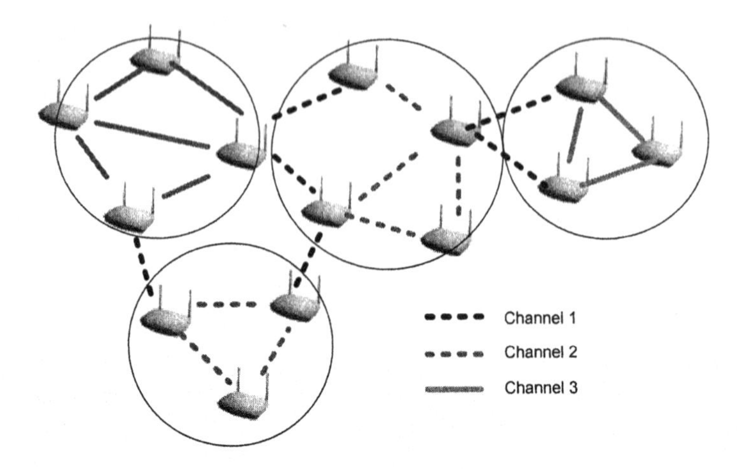

Figure 7.4. Clustering Multiple Channels Architecture with Two Radios

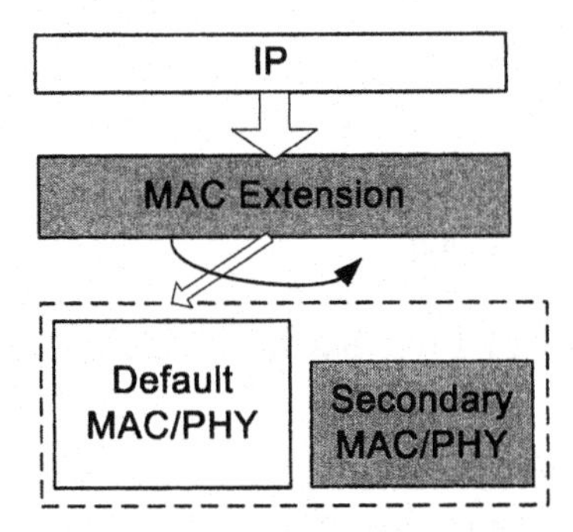

Figure 7.5. Two-Radio Protocol Stacks

- A node is elected as a clusterhead if it is the most highly connected (having the highest number of neighboring nodes) node of all its “uncovered" neighbor nodes (in case of a tie, lowest ID (e.g. MAC address) prevails);

- A node which has not elected its clusterhead is an “uncovered" node, otherwise it is a "covered" node;

- A node which has already elected another node as its clusterhead gives up its role as a clusterhead.

To minimize the co-channel interference (CCI) among clusters, we propose a Minimum Interference Channel Selection (MIX) algorithm, by which a clusterhead selects the secondary radio channel (denoted as k) with the minimum energy on air for intra-cluster communication.

Let $\bar{E}_i$ denote the average energy on the ith channel for the duration T, we have

$$\bar{E}_i = \frac{\int_{t=t_0}^{t_0+T} E_i(t)dt}{T}, \tag{7.16}$$

where $E_i(t)$ is the instantaneous energy on the ith channel at time t. Hence, the MIX algorithm is represented by

$$\{k \mid \bar{E}_k = \min(\bar{E}_i | i = \{1, 2, ..., n\})\}, \tag{7.17}$$

where n is the total number of orthogonal channels. Obviously, the longer the estimation duration T, the more accurate the estimation. Our simulations used $T = 2$ seconds.

A clusterhead will generate a pseudo random number with 6 bits length for the ID of its cluster. Also it is responsible for notifying all its members which channel is used to configure the secondary radio as well as when the channel information is expired (denoted as T_E (Eq.7.18)).

$$T_E = T_o + T_1 + \text{uniform}(0,\ T_2), \tag{7.18}$$

where T_o indicates the time when the clusterhead selected the channel, and T_1 and uniform$(0,\ T_2)$ [5] give the constant and random components of the lifetime, respectively. Our simulations used $T_1 = T_2 = 100$ seconds.

When the channel information is expired, the clusterhead will re-run the MIX algorithm to select a new channel, then broadcast the updated channel and its lifetime to its cluster members.

After getting the channel information, the neighboring nodes notify each other the channel used by their secondary radio. Thereby, we build a new 16-bit CMT field (see Fig.7.6) with three sub-fields: status, channel, and number of uncovered neighboring nodes. The "cluster-ID" flag indicates the cluster that the node belongs to, and is only meaningful when "status" is not "uncovered"; the "number of uncovered neighboring nodes" is used for electing clusterhead. In our OPNET implementation, the new 16-bit CMT field is added into the 802.11 DATA frame. The 16-bit "Duration ID" field in the legacy 802.11 ACK frame can also be used as the new CMT field, since it is meaningless when segmentation is not used or the ACK is for the last fragment of the packet.

After learning that a peer node belongs to the same cluster, a node will configure the forwarding table in its extended MAC such that all packets destined to the peer node go through the secondary MAC/PHY module. Fig.7.7 summarizes the above clustering and channel selecting procedures with a state transition diagram.

5. Simulation Results and Discussions

In this section, we present results from a series of simulations to demonstrate the effectiveness of physical carrier sensing with tunable sensing thresholds in

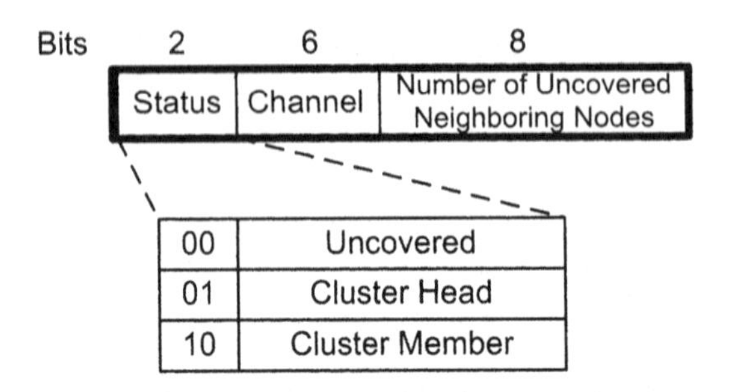

Figure 7.6. Definition of 16-Bit CMT Field

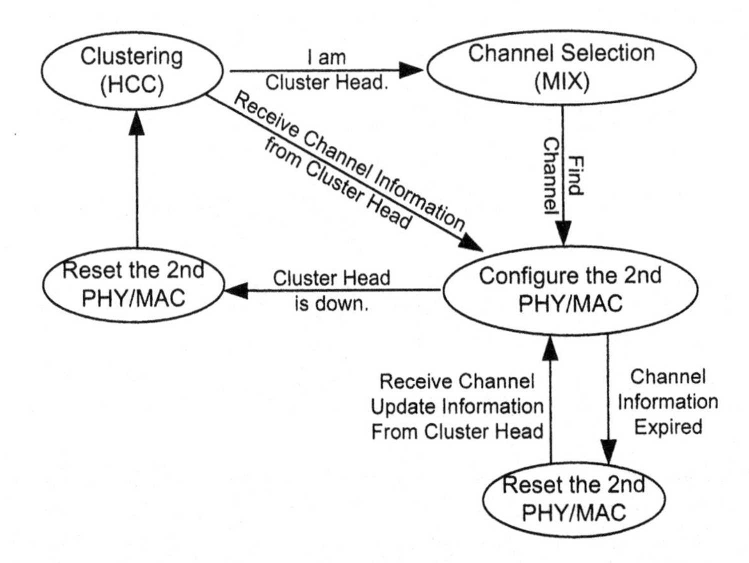

Figure 7.7. A State Transition Diagram of Clustering and Channel Selecting Procedures

improving network performance for various topologies. All the simulations were conducted in the OPNET simulation environment [15]. We have extended OPNET kernel modules to support tunable physical carrier sensing, a configurable propagation environment and multiple 802.11b data rates.

In all simulations, we configured each node to be always backlogged with 1024 bytes long MAC data frames. Each node transmits at a fixed power of 0 dBm. By default, the OPNET simulator configures the physical carrier sensing threshold to be the same as the reception threshold P_R. Furthermore, the ambient noise level was set at -200 dBm.

The primary performance metric studied in this paper is throughput, defined as the total number of bits successfully received in a second. As per the .11

MAC, if the sender does not receive an ACK for a transmitted data packet, it assumes that the data packet is lost and performs a retransmission. However, it is also possible that the data packet is received correctly, but the ACK is lost. This also causes a retransmission and can result in a multiple copies of the same data packet at the receiver. In this case, only the first one received will be forwarded up to higher layers, and the rest discarded. When computing throughput in this paper, we only count successful non-duplicate data packets , i.e. the *goodput* which underestimates the actual throughput on the physical channel of the network. However, since ACK packets are much shorter than data packets and they are typically transmitted using the lowest (most reliable) data rate in 802.11, the probability of successfully receiving a data packet but losing an ACK packet is very low. Thus, we assume that *throughput* is approximately equal to *goodput* in an 802.11 network.

Note that the .11 MAC employs contention management via the binary exponential backoff (BEB) with a configurable contention window (CW) size parameter. This random scheduling of user transmissions naturally impacts the received SINR in addition to the various mechanisms described in this paper. Since the primary focus of the simulations in this section is on interference avoidance via PCS, the BEB mechanism is disabled in our simulations and the contention window size is fixed at the maximum value for 802.11b (CW = 1024) to minimize the likelihood of collisions due to simultaneous transmission. This configuration allows us to isolate the specific effects of adaptive physical carrier sensing on network performance.

5.1 Point-to-point baseline performance of 802.11b MAC

To validate the effectiveness of physical carrier sensing, we need the following two baseline figures: the SINR thresholds (S_0) required to sustain each available data rate in an 802.11b network, and the effective MAC throughput at each data rate. In the first simulation, we configured a network of two nodes – one sender and one receiver. The pathloss exponent was set to 2 to reflect a free-space environment. With RTS/CTS disabled, we varied the T-R separation distance and measured the effective throughput provided by the MAC layer at the receiver. The same simulation sequence was repeated for all four data rates defined in the 802.11b standard.

The results are shown in Fig.7.8 where instead of the T-R distance, the throughput is shown against the SINR at receiver. Hence the results depict the fundamental relationship between MAC throughput and receiver SINR. This mapping is valid irrespective of pathloss, transmission power and T-R distance. These results, recorded in Table 7.2, will be used to design and

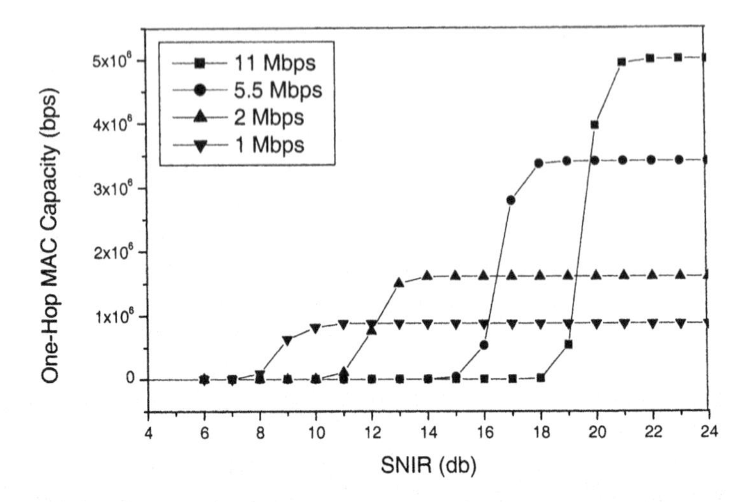

Figure 7.8. One-Hop multi-rate performance of 802.11b for various SINR values at the receiver (RTS/CTS disabled).

Table 7.2. One-hop performance of 802.11b MAC without RTS/CTS

Data Rate (Mbps)	1	2	5.5	11
S_0 (dB)	11	14	18	21
W (Mbps)	0.89	1.5	3.5	5.0

analyze simulations in the rest of the section. The results confirm that MAC overhead is generally larger at higher data rates and higher data rates require higher SINR thresholds, as expected.

5.2 Maximizing Spatial Reuse with the Optimal PCS

We conduct simulations in two scenarios with regular topology: 90-node chain and 10×10 grid. The goal is to validate the theoretical optimal PCS threshold β, derived in Section 3.2.

First, we expanded the previous network into a chain of 90 nodes (to approximate an infinite chain). The only traffic allowed is originated by node 1 and designated for node 90, with the other 88 intermediate nodes acting as relays. The reception power threshold (P_R) was configured such that the transmission range is 13 meters. Each node relied on physical carrier sensing only to avoid interference using identical carrier sensing threshold and data rates. We measured the end-to-end throughput while varying the sensing threshold and the data rate. The results for $\gamma = 2$ are plotted in Fig. 7.9.

Note that the results show the existence of an optimal sensing threshold value for each data rate. With everything else fixed, altering the data rate changes the SINR requirement (S_0), hence the optimal sensing threshold changes as well. Also notice that the common practice of having the carrier sense threshold equal to the reception threshold, i.e., $P_{cs\_t} = 0db$ corresponds to the right-most point on respective curves. Hence, the throughput improvement achieved by tunable physical carrier sensing threshold can be as high as 4 times (at data rate 11 Mbps).

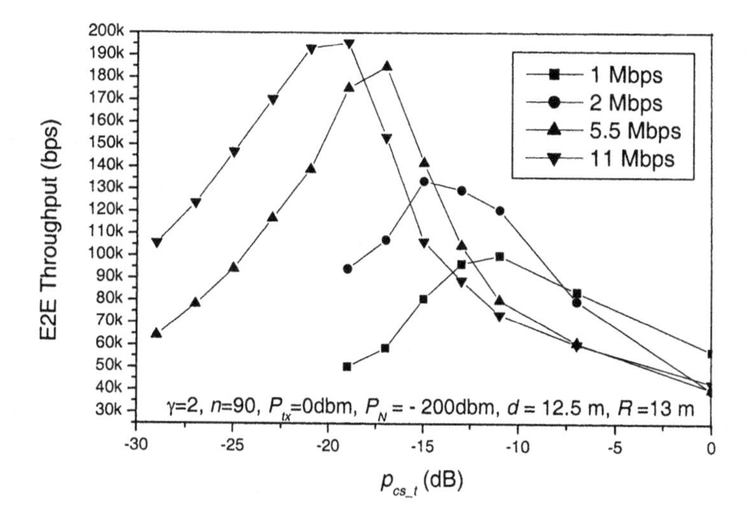

Figure 7.9. End-to-end throughput in a 90-hop chain for various sensing thresholds and data rates.

Table 7.3 compares the optimal sensing threshold p_{cs_t} obtained from the simulations against the theoretical optimum β. As the table shows, the two values matches very well.

Table 7.3. Optimal carrier sensing thresholds (dB) in a 90-node chain ($\gamma = 2$)

Data Rate (Mbps)	1	2	5.5	11
Simulation	-11	-15	-17	-19
β	-11	-14	-18	-21

Table 7.4 compares the optimal throughputs obtained from the simulations against the prediction from the spatial reuse study described in Section 3.3. The theoretical prediction assumed a perfect MAC protocol that always derives the globally optimal schedule for communications; this yields a theoretical upper-bound of network throughput. As shown in Table 7.4, the network with

optimally tuned physical carrier sensing was able to achieve around 90% of the theoretical maximum.

Table 7.4. Optimal E2E throughput in a 90-node chain

Data Rate (Mbps)	1	2	5.5	11
W (Mbps)	0.89	1.5	3.4	5.0
k (spatial reuse)	7.1	10	15.9	22.4
T: Theoretical (W/k)	0.105	0.15	0.21	0.223
S: Simulation	0.1	0.134	0.185	0.196
S/T	95%	89%	88%	88%

Next, we turn to a 2-D network: 10×10 grid, which is more representative of typical real world topologies. Each packet has its own destination chosen randomly from the immediate neighbors of the transmitter. In this configuration, the Manhattan distance between neighboring nodes was 4.5 meters. The reception power threshold (P_R) was configured to allow the transmission range of only 4.5 meters such that only immediate neighbors could directly communicate.

We conducted four sets of simulations using 1 Mbps, 2 Mbps, 5.5 Mbps and 11 Mbps as the data rate for each node, respectively. In each set of the simulations, we altered the path loss exponent and PCS threshold. The aggregate throughput of the grid network are plotted in Fig. 7.10. It is evident that the optimal PCS threshold does not change with the path loss exponent in a large homogeneous network, and the optimal PCS threshold obtained via simulation matches the theoretical β very well (see Table 7.5).

Table 7.5. Simulation results of optimal carrier sensing thresholds (dB) in a 10×10 802.11b grid (S: Simulation)

Data Rate (Mbps)	1	2	5.5	11
Simulation ($\gamma = 2$)	-11	-13	-17	-19
Simulation ($\gamma = 2.5$)	-11	-13	-17	-20
Simulation ($\gamma = 3$)	-11	-13	-17	-20
β	-11	-14	-18	-21

The simulation is now repeated for 802.11a [6]. Table 7.6 compares the theoretical optimal p_{cs_t} (i.e. β) with the optimal value from simulations, showing that the theoretical optimal carrier sensing threshold β is also valid for 802.11a network.

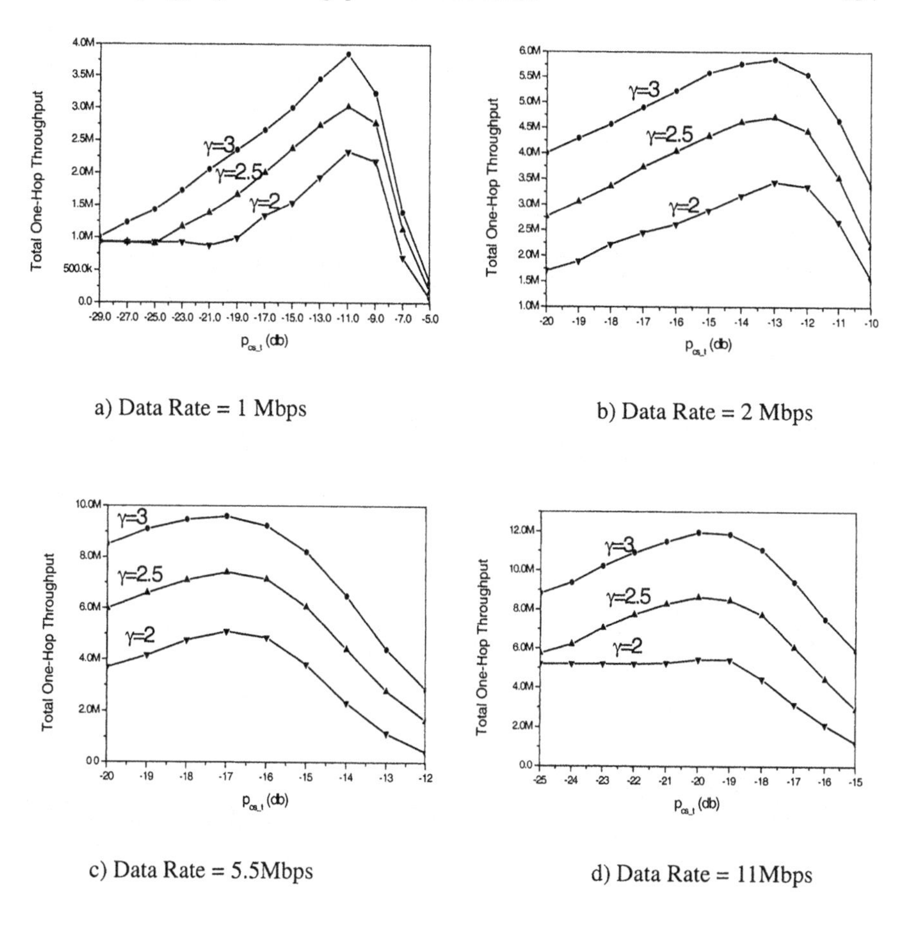

Figure 7.10. Aggregate 1-hop throughput as a function of PCS threshold for various pathloss exponent values in a 10 × 10 802.11b grid (data rate =1Mbps, n=10x10, Random Traffic, P_N=-200dbm)

Table 7.6. Optimal carrier sensing thresholds (dB) in a 10 × 10 802.11a grid with Different Data Rate (Mbps) ($\gamma = 3$)

Data Rate (Mbps)	6	9	12	18	24	36	48	54
Simulation	-7	-9	-11	-13	-17	-21	-27	-29
β	-7	-9	-11	-13	-17	-22	-27	-29

5.3 Optimal PCS + Multi-Channel Clustering

Fig. 7.11 shows an example of how our clustering multi-channel and two-radio architecture works in a 10×10 regular grid with three orthogonal channels. We implement a two-radio 802.11 client module in OPNET. Let d denote the distance between two nearest neighbors; we configure the transmission range

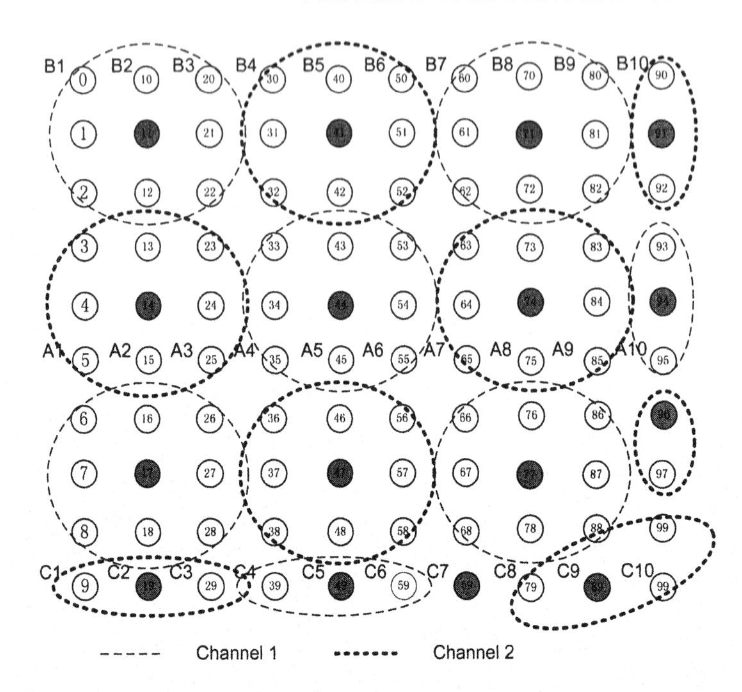

Figure 7.11. An Example of 10 × 10 Grid Using the Clustering Multi-Channel and Two-Radio Architecture

as $\sqrt{2}d$. During the simulation, the network will automatically cluster into the topology as shown in Fig.7.11. The dark nodes are clusterhead, and the dotted circle indicates all independent clusters. Two orthogonal channels (for the circles denoted by thick and thin dash lines, respectively) are used for intra-cluster communications. The channel assignment in Fig.7.11 minimizes co-channel interference and achieves the highest spatial reuse.

Fig.7.12 compares the total one-hop throughput with the new clustering multi-channel and two-radio architecture to the traditional single-channel, single-radio mesh. A random traffic generation model at each node was used with a sufficiently high offered load such that the nodes remain saturated during the simulation. Fig.7.12 clearly demonstrates the performance improvement with clustering and multiple orthogonal channels. The steady-state average throughput is 8.1 Mbps in the new CMT architecture, and only 2.7 Mbps in the single-channel and single-radio mesh. The gain is about 300%, which is the maximum gain achievable when using 3 orthogonal channels.

Fig.7.13 illustrates the one-hop throughput distribution with respect to links, where links A_i, B_i, and C_i are also indicated in Fig. 7.11 ($i = \{1, 2, ..., 10\}$). Clearly, links A_i experience worse interference environment than links B_i and C_i, leading to the oscillation of the throughput distribution, illustrating the

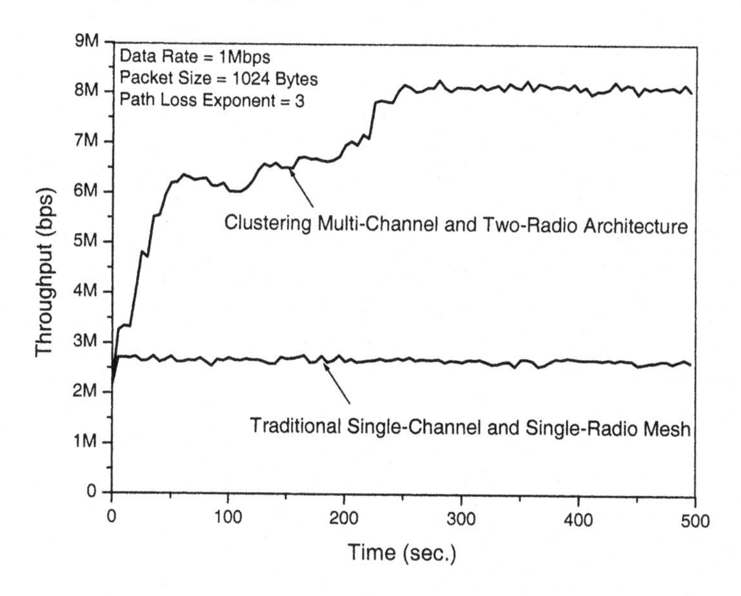

Figure 7.12. Total One-Hop Throughput Comparison

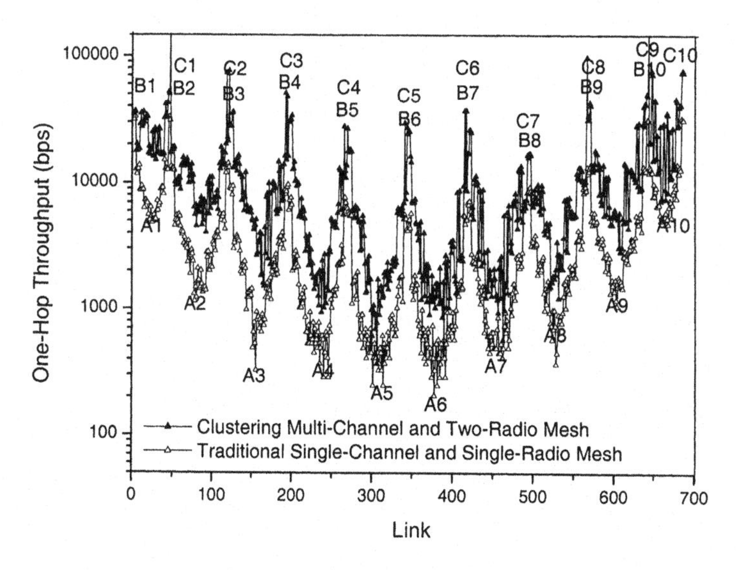

Figure 7.13. Average One-Hop Throughput Distribution (between 300sec. and 500 sec.)

location-dependent fairness problem. We do not consider the fairness problem further; it is interesting to speculate how physical carrier sensing may be used to mitigate the *location-dependent fairness* problem. It implies for instance that the preferable locations for gateways in a wireless mesh network may not be the center of the network.

Finally, we validate the new architecture in a random topology as shown in Fig.7.14. The transmission range is fixed at 25 meters. In Fig.7.14, dash and dash-dotted lines indicate the nodes using channel 1 and channel 2 for their secondary radio, respectively. We also used the circles to illustrate the clusters with clusterhead in the center. Fig.7.15 compares the performance with both aggregate throughput and throughput distribution. We clearly see that (after 300 seconds) the aggregate throughput of the proposed architecture (10Mbps) is almost 3 times higher than that of the traditional one (3.5Mbps).

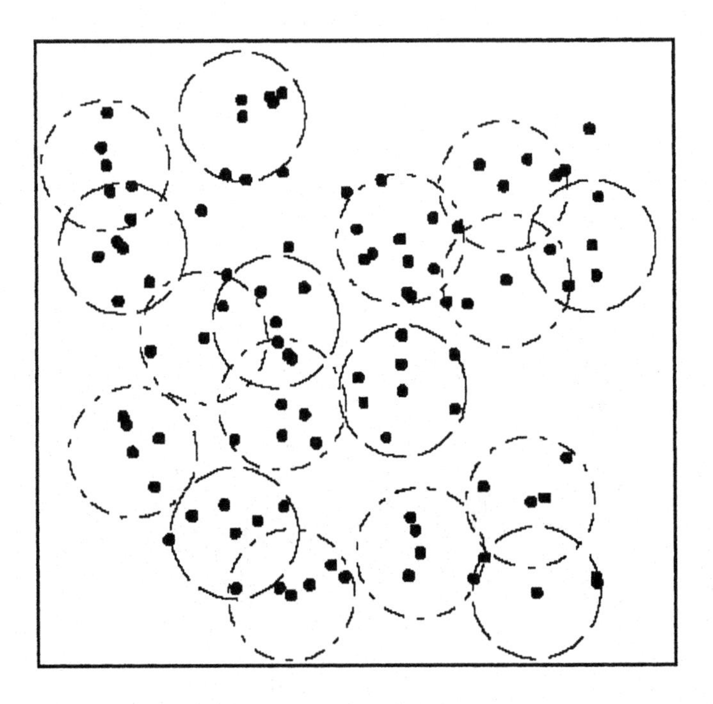

Figure 7.14. 2D 200m x 200m 100-Nodes Random Topology Using the Clustering Multi-Channel and Two-Radio Architecture (at 500 seconds)

6. Related Work on .11 MAC Enhancements

Interference mitigation has been a well-known challenge for MAC protocols in wireless mesh networks. Much of the existing research in this space has focused on eliminating the *hidden terminal* [11] problem. A virtual carrier

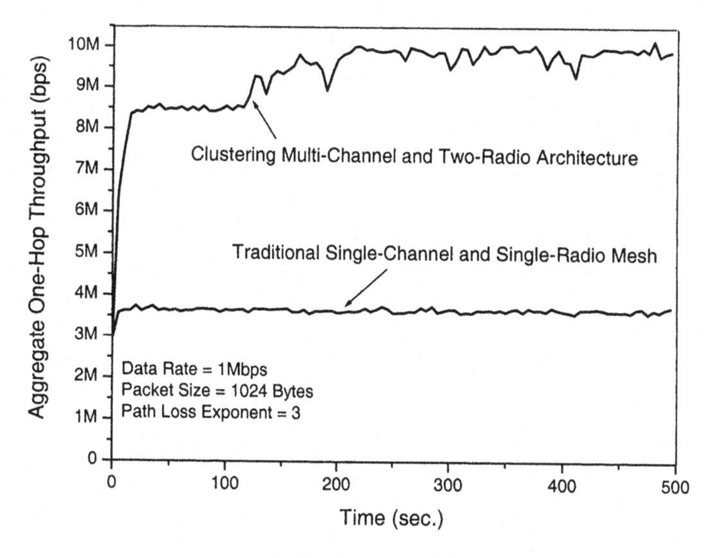

a) Tracing Aggregate Throughput

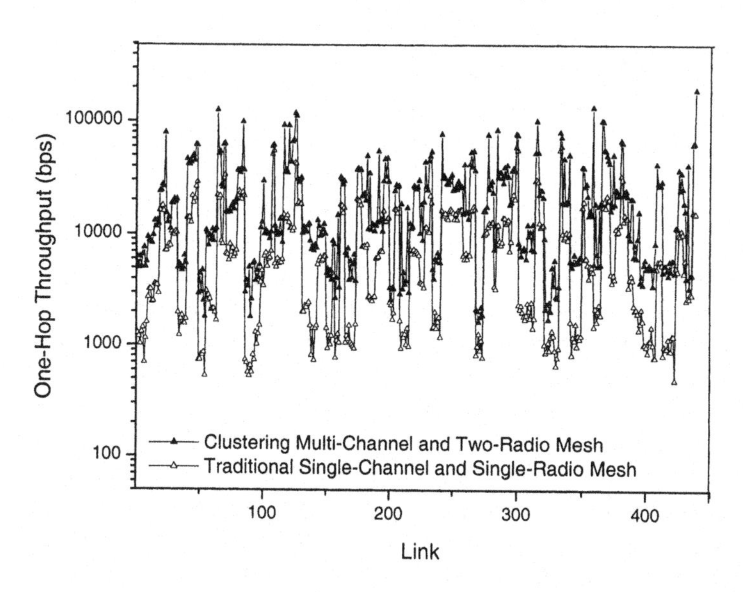

b) Throughput Distribution

Figure 7.15. Performance Comparison in the Random Topology

sensing mechanism, implemented through the RTS/CTS handshake, has been adopted by IEEE 802.11 in an attempt to eliminate the hidden terminal problem. However, this has an underlying assumption that all hidden terminals are within transmission range of receivers (allowing them to receive the RTS or CTS packet successfully). While such an assumption may be reasonable for single cell WLANs, it is generally not true for multi-cell WLANs and multi-hop mesh networks. Researchers [12] [13] [14] have by now recognized that virtual carrier sensing with RTS/CTS does not solve the hidden terminal problem effectively for such networks.

It was shown in both [12] and [14] that the interference range is a function of T-R separation distance. Depending on the T-R separation distance, the interference range can be smaller or larger than the transmission range. If the interference range is smaller than the transmission range, RTS/CTS can indeed prohibit all the hidden terminals from interfering with the existing transmission; but some of the nodes that are not capable of interfering are also prohibited from transmitting. Thus, in this configuration RTS/CTS is too aggressive, resulting in a significant exposed terminal problem that wastes potential throughput by requiring potential transmitters to unnecessarily back off. On the other hand, if the interference range is larger than the transmission range, RTS/CTS can fail to prevent hidden terminals from interfering with an existing transmission. So RTS/CTS is too conservative and ineffective in this case.

A technique was suggested in [12] to avoid the conservative RTS/CTS scenarios by allowing only the transmitter-receiver pairs with distance shorter than a threshold to perform transmission; the threshold is set such that the corresponding interference range will not be larger than the transmission range. The constraint on T-R separation distance is imposed by only allowing a node to reply to a RTS packet with a CTS packet when the receive power of the RTS packet is larger than a threshold, even if the RTS packet is received successfully and the node is idle. This added constraint ensures that RTS/CTS never becomes too conservative and so the hidden terminal problem is avoided. However, this approach does not address the exposed terminal problem introduced by the aggressive RTS/CTS. Another disadvantage of such an approach is that it reduces effective transmission range and thus lowers network connectivity.

Several other techniques attempt to reduce inefficiencies introduced by exposed terminals. The protocol described in [10] focuses on the exposed terminal problem directly by enabling nodes to identify themselves as exposed nodes and opportunistically scheduling concurrent transmissions whenever possible. While [14] recognizes that RTS/CTS can be either too conservative or too aggressive, it only addresses the problems associated with aggressive RTS/CTS. The authors propose a Distance-Aware Carrier Sensing (DACS) scheme which employs an extra handshake in addition to RTS/CTS to disseminate one-hop distance information to neighbors so that medium reservation can be more accurate and spatial reuse can be improved to reduce the negative impact of exposed terminals.

Besides interference, packet collision is another important factor contributing to the final SINR. However, collisions due to simultaneous transmission attempts cannot be reliably prevented by using physical carrier sensing alone. A common approach to avoiding persistent collisions is random back-off, e.g the binary exponential backoff algorithm in the 802.11 MAC standard. Recently, there have been several efforts aimed at optimizing the back-off algorithms and contention window size to minimize collisions [6] [17] etc. While the focus of this paper is on leveraging the spatial reuse of a network to enhance the throughput performance through physical carrier sensing, these techniques may be supplemented with the above to simultaneously minimize interference and the impact of packet collisions to further improve aggregate throughput in a dense wireless network.

Unlike prior techniques that attempt to avoid interference through handshake protocols, this paper approaches interference mitigation from the perspective of leveraging spatial reuse. We believe that the key to the optimal spatial reuse is to maintain the appropriate separation distance between simultaneous transmitters. Therefore we focus on enhancing the physical carrier sensing mechanism with tunable sensing threshold for the 802.11 MAC. What we propose in this paper is a simple and effective method that directly redresses some of the issues in virtual carrier sensing with RTS/CTS.

7. Conclusion

In this paper, we propose to enhance physical carrier sensing with a dynamically tunable sensing threshold and adopt a novel clustering two-radio and multi-channel architecture to improve spatial reuse in 802.11 mesh networks, aiming at increasing the aggregate network throughput. Simulations were performed for both 1-D chain and 2-D grid topologies to validate the analysis and the proposed scheme. The main contributions of this paper are:

(1) We have demonstrated that physical carrier sensing with the tunable sensing threshold is effective at leveraging spatial reuse in 802.11 multi-hop mesh networks, shown by increases in aggregate throughput. This improvement is achieved without requiring the use of virtual carrier sensing. Although the 802.11 MAC is a CSMA/CA based distributed and asynchronous scheme, it has the capability to make good use of the spatial-reuse property in a mesh (90% of the theoretical limit in a chain).

(2) We have proposed an adaptive PCS scheme to achieve a near-optimal carrier sense threshold automatically, leading to a substantial throughput improvement for 802.11 mesh networks. With assumptions of homogeneous links and co-location of sender and receiver, $1/S_0$ gives the theoretical approximation to the optimal sensing threshold, where S_0 is the SINR threshold for achieving the link capacity.

(3) We have presented a new clustering multi-channel and two-radio (CMT) architecture using 802.11 MAC protocols. Distributed clustering works with a new minimum interference channel selection algorithm (MIX) to distribute orthogonal channels in a mesh, maximizing the aggregate throughput. OPNET simulations were conducted to validate the new architecture. Compared to a traditional single-channel and single-radio mesh, the gain achieved with three orthogonal channels in terms of the aggregate one-hop throughput is about 300% in a 10×10 grid using local, random, and saturate traffic as well as in a 200m x 200m 100-nodes random topology.

Although this paper is focused on 802.11 networks, the analysis on optimal physical carrier sensing is applicable to any CSMA/CA-based mesh network. As the initial step to showcase the potential of enhanced physical carrier sensing and multi-channel clustering in improving aggregate throughput, this paper focuses on regular network topologies. Future work may include extending the investigation to random topologies.

Notes

1. Further, proportionally improved MAC efficiency is needed to translate increase link layer rates to higher MAC throughput.
2. Note that each AP also needs a third radio for communications for the AP-MT link (MT: Mobile Terminal), which must be orthogonal to the AP-AP mesh band to avoid interference. This promotes the use of dual, e.g. .11 a/b, radio interface cards.
3. Clearly, this assumption is a main drawback of the subsequent analysis, but further refinements are not possible without assuming specific network topologies. The results here thus have the advantage of not being tied to a specific topology.
4. More so for higher data rates since higher S_0 values will be required.
5. a random variable with uniform distribution on the range $(0, T_2)$; the random component is designed to avoid the event that two clusters always select the channel at the same time, i.e., channel selection collisions.
6. We use the same modulation curve for 802.11a simulation as in [18]

References

[1] IEEE Standard for Wireless LAN Medium Access Control (MAC) and Physical Layer (PHY) specifications, ISO/IEC 8802-11: 1999(E), Aug. 1999.

[2] MeshDynamics, http://www.meshdynamics.com.

[3] MeshNetworks, http://www.meshnetworks.com.

[4] Packethop, http://www.packethop.com

[5] B. P. Crow, J. G. Kim, "IEEE 802.11 Wireless Local Area Networks," IEEE Comm. Mag., Sept. 1999.

[6] G. Bianchi, "Performance Analysis of the IEEE 802.11 Distributed Coordination Function," IEEE JSAC, vol. 18, no. 3, March 2000.

[7] X. Guo, S. Roy, W. Steven Conner, "Spatial Reuse in Wireless Ad-Hoc Networks," VTC2003.

[8] X. Guo, "Personal research notes," 2003.

[9] A. Adya, P. Bahl, J. Padhye, A. Wolman and L. Zhou, "A Multi-radio Unification Protocol for IEEE 802.11 Wireless Networks," Tech. Rpt. MSR-TR-2003-44, July 2003.

[10] D. Shukla, L. Chandran-Wadia, S. Iyer, "Mitigating the exposed node problem in IEEE 802.11 adhoc networks," IEEE ICCCN 2003, Dallas, Oct 2003.

[11] F. A. Tobagi, L. Kleinrock, "Packet Switching in Radio Channels: PART II- The Hidden Terminal Problem in Carrier Sensing Multiple Access and Busy Tone Solution," IEEE Trans. on Commun, Vol. COM-23, No. 12, pp. 1417-1433, 1975.

[12] K. Xu, M. Gerla, S. Bae, "How effective is the IEEE 802.11 RTS/CTS handshake in ad hoc networks?" GLOBECOM02, Nov 17-21, 2002.

[13] S. Xu, T. Saadawi, "Does the IEEE 802.11 MAC Protocol Work Well in Multihop Wireless Ad Hoc Networks?" IEEE Communications Magazine, P130-137, June 2001.

[14] F. Ye, B. Sikdar, "Improving Spatial Reuse of IEEE 802.11 Based Ad Hoc Networks," To appear in the Proceedings of IEEE GLOBECOM, San Francisco, December 2003.

[15] http://www.opnet.com

[16] Intersil, "Direct Sequence Spread Spectrum Baseband Processor," Doc# FN 4816.2, Feb. 2002, http://www.intersil.com/

[17] V.Bharghavan, A. Demers, S. Shenker, and L. Zhang, "MACAW: A Media Access Protocol for Wireless LANs," in Proc. Of ACM SIGCOMM'94,

[18] A.J. van der Vegt, "Auto rate fallback algorithm for the ieee 802.11a standard," http://www.phys.uu.nl/ vdvegt/docs/gron/.

[19] Theodore S. Rappaport. (2002). *Wireless Communications, Principles and Practices*. 2nd Ed. Prentice Hall.

[20] J. Monks, V. Bharghavan, W. Hwu, "A Power Controlled Multiple Access Protocol for Wireless Packet Networks," in IEEE INFOCOM, April 2001.

[21] R. Roy Cloudhury, X. Yang, R. Ramanathan and N. H. Vaidya, "Using Directional Antenna for Medium Access Control in Ad Hoc Networks," in ACM MOBICOM, Atlanta, GA, Sept. 2002.

[22] M. Gerla and J.T.-C. Tsai, "Multicluster, mobile, multimedia radio network", ACM/Baltzer Journal of Wireless Networks. vol. 1, (no. 3), 1995, p. 255-265.

[23] C. R. Lin, M. Gerla, "Adaptive Clustering for Mobile Wireless Networks," IEEE Jour. Selected Areas in Communications, pp. 1265-1275, Sept. 1997.

[24] Y. P. Chen and A. L. Liestman, " Zonal Algorithm for Clustering Ad Hoc Networks," International Journal of Foundations of Computer Science, 2003.

[25] N. Jain and S. Das, "A Mutlichannel CSMA MAC Protocol with Receiver-Based Channel Selection for Mutlihop Wireless Networks," in Proceedings of the 9th Int. Conf. on Computer Communications and Networks (IC3N), Oct. 2001.

[26] S.-L. Wu, C.-Y. Lin, Y.-C. Tseng and J.-P. Sheu, "A New Mutli-Channel MAC Protocol with On-Demand Channel Assignment for Multi-Hop Mobile Ad Hoc Networks, " in IEEE Wireless Communications and Networking Conference (WCNC), Chicago, IL, Sept. 2000.

Chapter 8

A SELF-CONFIGURING LOCATION DISCOVERY SYSTEM FOR SMART ENVIRONMENTS

Andreas Savvides
Electrical Engineering Department
Yale University, New Haven, CT 06511
andreas.savvides@yale.edu

Mani Srivastava
Electrical Engineering Department
University of California, Los Angeles, CA 90095
mbs@ee.ucla.edu

Abstract This chapter describes our experiences with the design and use of a rapidly installable self-configuring beaconing system. Our system is comprised of a set of custom-designed wireless sensor nodes, the Medusa MK-2 nodes that can form a local coordinate system a few seconds after they are deployed. This results in a versatile, low-cost system based on battery operated devices eliminates the need for costly and time-consuming infrastructure installations. During the system bootstrapping phase, the Medusa Kindergarten - 2(MK-2) nodes localize themselves and then enter a service phase in which they act as "satellites" that assist other objects in the room to determine their locations with a few centimeters of accuracy.

Keywords: Sensor networks, node localization, location aware systems.

Introduction

As wireless embedded systems evolve, location awareness is becoming a fundamental requirement for small wireless devices to operate autonomously and it is the key enabler for many applications. This chapter reports our experiences in adding location awareness to a deeply instrumented system for observing the level of children interaction during early childhood education. This work was conducted as part of the Smart Kindergarten project [9, 8] at the

University of California, Los Angeles. The Smart Kindergarten project aims to develop a deeply instrumented environment for studying child development in early childhood education. The overall goal of the project is to provide the necessary infrastructure support for investigating how learning takes place inside a classroom during a child's pre-school years. To do so the members of the Smart Kindergarten team have designed and implemented a wide variety of platforms and components that enable the close observation of student behavior in the form of speech, gestures and their interaction with toys and objects in a classroom setup. The Smart Kindergarten environment hosts a wide variety of devices that ubiquitously record the activities of children inside the classroom. A *Smart Table* [3] was developed to track multiple objects sitting on its surface. By identifying and localizing various objects on the table surface, the Smart Table measures the level of interaction of kids with a set of puzzle blocks sitting on the table. An elaborate speech processing system has been designed to record and process speech streams, and a rapidly deployable ad-hoc localization system has been designed to estimate and record high precision positions of students (within 10 centimeters) and objects within the classroom. This time-stamped information is propagated and stored in a backend database that can be used by the education researchers that need to process the data offline at a later stage.

This chapter focuses on the localization subsystem components. The uniqueness of this system lies in the new innovative design of an ecology of sensing devices, supported by a software infrastructure and a new set of location discovery algorithms specifically designed for this project. Instead of providing all the details of the project, this chapter highlights the main features of the smart beaconing system, and provides a set of references that describe each component in greater detail. The first half of the chapter describes the software and hardware components of the localization infrastructure, and the second half presents an overview of our results in the broader topic of ad-hoc node localization. The chapter concludes with summary of the lessons learned and a set of directions for future work.

1. Localization System Hardware Building Blocks

To closely observe student locations inside a classroom we have developed a location infrastructure shown in Figure 8.1. The infrastructure based on an ecology of wirelessly connected sensing devices that among other tasks are responsible for extracting detailed location information from the classroom environment. The students are tracked with the help of a custom designed wearable device, the *iBadge* [2], that is able to obtain its location with the help of a set of smart beacons, the Medusa MK-2 beacon nodes attached on the classroom ceiling. Other objects in the room are tracked with an object

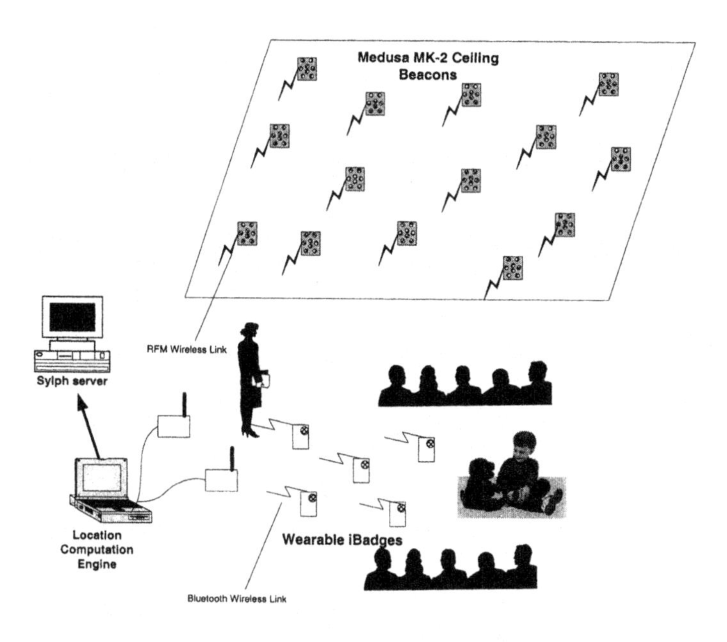

Figure 8.1. The Smart Kindergarten Localization Infrastructure

tag which is implemented with a Mica wireless sensor node designed at UC Berkeley. All the devices are battery operated and the system is designed to be rapidly deployable and self-configuring. Operation in a typical classroom setting will proceed as follows. First the ceiling beacons are evenly placed on the classroom ceiling. During an initial *bootstrapping phase*, the ceiling beacons form a local coordinate system by measuring the horizontal distances to each other using their onboard ultrasonic distance measurement system. This process takes a few seconds to complete and the locations of the ceiling beacons are stored on a workstation that serves as the *location computation engine*. Once the bootstrapping phase is completed, the ceiling beacons enter a *service mode*. When in service mode, the beacons synchronize among themselves to broadcast a combination of radio and ultrasound reference signals into the classroom space at a frequency of approximately 12 reference signals per second. The iBadges and object tags, use these signals to compute their distances to the beacons (the Medusa MK-2 nodes) and transmit their time-stamped distance measurements to the location computation engine that computes node locations and stores them in the Sylph server [10] for future processing.

While it is possible to pursue a distributed implementation where beacons could transmit their signaling in a more ad-hoc manner and devices in the room could estimate their locations individually, we decided on a centralized

Figure 8.2. The Smart Kindergarten Localization Platforms

implementation for performance purposes. By having beacons transmit their reference signals in a synchronous manner, the localization process can achieve a higher rate, that is, devices inside the room can receive location updates at a faster rate. For the same reason, the devices inside the classroom propagate their measurements to the location computation engine, to achieve higher rate processing of the measurements. In addition to performance issues, this setup was also selected to favor scalability. By having a limited set of ceiling beacons transmitting the reference information, and allowing wearable devices and object tags to make passive measurements, one can accommodate a very large number of devices in the same room without compromising the location update rate of the system. Figure 8.2 depicts the platforms designed for this system. A more detailed description of the platforms is provided in the subsections that follow.

1.1 Ceiling Beacons: *Medusa MK-2*

The Medusa MK-2 node [5] is a low cost, low power wireless sensor node, specifically designed to act as a ceiling beacon node. The MK-2 node has two on board microcontrollers, an ATMega128L and an AT91FR4081, both from ATMEL. The ATMega128L is an 8-bit microcontroller running at 4MHz and it is responsible for running frequent tasks that do not require any heavy weight computation. Such tasks include driving the on-board RFM TR1000 radio,

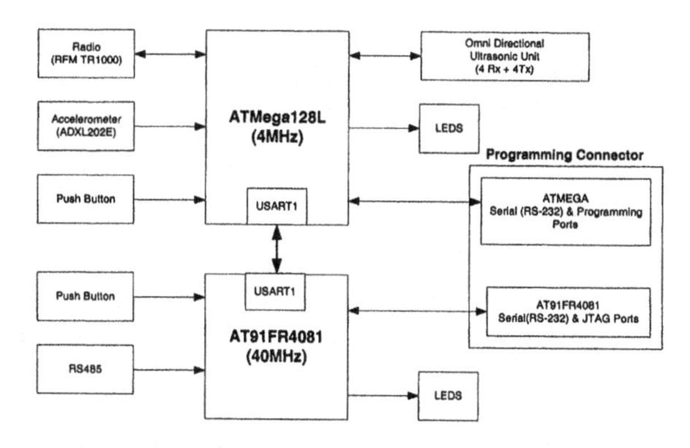

Figure 8.3. Medusa MK-2 architecture

handling a lightweight MAC layer and driving the ultrasonic ranging system. The different software components are implemented as tasks inside the PALOS embedded operating system that will be described in section 2.1.

The AT91FR4081 processor features an ARM7TDMI core supporting the 32-bit ARM and the 16-bit THUMB instruction sets. This processor runs at 40MHZ and has 128Kbytes of on-chip RAM and 1Mbyte of FLASH. This processor is dedicated to less frequent but more computation intensive tasks such as the least squares estimation algorithms used to estimate node locations during the bootstrapping phase. The location computation in a distributed fashion inside the sensor node or it can take place on the sensor node using the algorithm described in [6] or at the central location computation engine. Figure 8.3 shows a block diagram that describes the MK-2 architecture.

In addition to the ultrasonic ranging board, the Medusa MK-2 node includes a 2-axis MEMS accelerometer to detect node movement, two push-buttons that can serve as user interfaces, as well as GPS interface to receive position and timing information from an additional outdoor node placed outside the classroom. The two processors communicate with each other through a UART and each processor has an additional UART attached to the programming connector pins that allows the processors to communicate with external devices.

The Ultrasonic Ranging Subsystem. The ultrasonic ranging system is a separate accessory board attached to the Medusa MK-2 node. This board carries four receiver-transmitter pairs of 40KHz, 120-degree beam angle ultrasonic transducers. Three transducer pairs are aligned in a circular pattern on the board perimeter, to provide a 360-degree angle. These transducers are slightly bent so that the ultrasonic signal transmissions coverage starts from the plane facing the node enclosure base and expands towards the plane perpendicular to the node base. The fourth receiver transmitter pair is perpendicular to the board to cover the region directly below the node. This transducer alignment produces a hemispherical coverage pattern that allows the ceiling beacons to measure distances to each other on the horizontal ceiling plane. At the same time, the ultrasonic transmissions are also transmitted in the room space, allowing the indoor localization of objects in 3D space. Each ultrasonic transmitter is driven by a separate general purpose I/O line from the ATMega128L microcontroller. The receiver circuit uses a two-stage op-amp amplifier followed by a threshold comparator and outputs a digital signal upon the detection of an ultrasonic signal. Each receiver output is wired to an external interrupt line that interrupts the ATMega128L microcontroller each time an ultrasonic transmission is detected.

Distance is measured by recording the time difference between the arrival of a radio pulse and an ultrasound pulse. The effective measurement range can range from a few centimeters to approximately 20 meters. In the MK-2 implementation, the maximum measurement range is about 4 meters. We found this range to be suitable for localizing objects in a room for several reasons. First, the transmission range is long enough to reach sensor nodes lying on the room floor. Since the transducers beam pattern has a lobe shape, the maximum horizontal range is slightly less (around 2.5 - 3 meters). This allowed the experimentation with the position calibration of smart beacons by using multihop measurements.

One of the main design parameters of the smart beacon system is multihop operation. This decision is based on four main reasons. First, the ability of the system to operate using multihop measurements, can help reduce localization latency. When low power ultrasonic transmissions are used, the beacons have to wait for a smaller interval for a any reverberations to die out before transmitting a new signal. Second, lower power ultrasonic transmissions; prolong the battery lifetime of each beacon. Third, multihop operation makes the system more scalable. This is desirable when deploying the ceiling beacons in rooms or over corridors in there is no full measurement connectivity between beacons. Multihop operation can localize beacons in the presence of obstacles by using indirect line-of-sight measurements.

1.2 Wearable units and object tags: *iBadge* and *Mica Ranger*

A wearable wireless badge, the iBadge [2], is equipped with a speech-processing unit, two radios (a Bluetooth and a low power RFM radio from RF Monolithics), 3D axis magnetic and acceleration sensors, temperature and humidity sensors and an ultrasonic ranging using to assist with localization of the iBadge. All the sensor measurements are transmitted through a Bluetooth link to a backend server, where all the data is stored in the Sylph middleware infrastructure. Sylph supports a set of abstractions that allow the easy connection of the sensors to the infrastructure and provides a suitable API that allows researchers to perform data mining on the collected data.

The object tag is implemented using a Mica sensor node from Crossbow [11]. It carries a custom design ultrasonic ranging board that is compatible with the ultrasonic ranging system on the Medusa MK-2 and iBadge nodes. The main difference of the object tag to the iBadge is that it can also act as a beacon by featuring both an ultrasonic transmission and reception. This was specifically implemented to facilitate localization in the presence of obstructions inside a room. The placement of object tags at different places inside the classroom, we can increase the probability that other objects and badges in the room to can be localized. This is because the object tags can transmit beacon signals from other locations in the room, to reach devices that have limited line-of-sight to the ceiling beacons.

2. Software Infrastructure Support

The localization infrastructure is supported by three main software components, the Palos embedded operating system running on all the wireless platforms (MK-2, iBadge and Mica Ranger), the measurement protocol stack that drives the ultrasonic measurement system and the location computation engine that fuses distance measurements to estimate node locations.

2.1 Palos

Palos (for Power Aware Lightweight OS) is a lightweight pseudo real-time non preemptive OS developed by Sung Park at the Networked and Embedded Systems Lab at UCLA to support the Smart Kindergarten platforms. Palos implements a lightweight pseudo real-time multitasking scheme where different functionalities are specified as a set of well defined tasks. The operating system consists of three main entities, the OS Core the hardware abstraction layer(HAL), a manager and a set of tasks. The HAL implements the drivers to microcontroller specific peripherals (e.g UART and SPI drivers) and platform specific components such the RFM radio drivers and the drivers for the ultrasonic ranging subsystems. Palos tasks are similar to threads on legacy operating systems. Tasks are registered with the OS core during system initialization and are either scheduled to execute periodically according to a set of user defined intervals. Each task can communicate with other tasks by depositing an event in each others event queue. Finally, the OS core runs the main control loop that determines the order and frequency of task execution and maintains the task event queues. When a task has turn to execute, any events previously deposited in the task queue are sequentially passed to the task. The operating system does not support task preemption but provides a set of mechanisms for prioritizing,

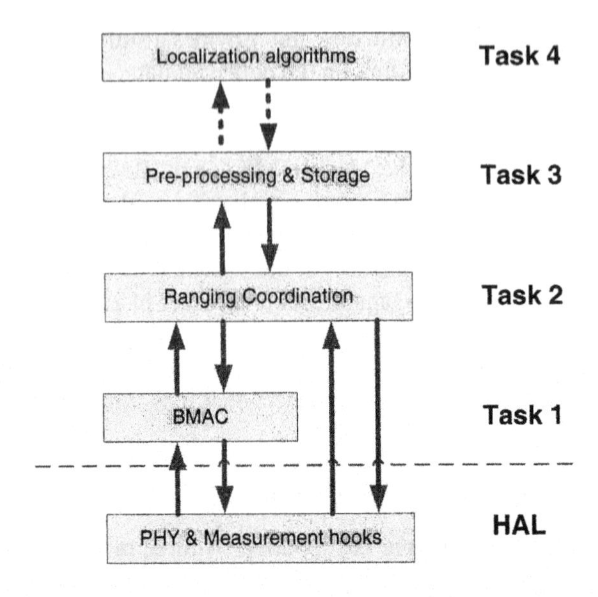

Figure 8.4. Measurement Protocol Stack

stopping and resuming tasks. The stripped Palos core of the operating system occupies 956 bytes of microcontroller FLASH and 548 bytes of RAM.

2.2 Distance Measurement Stack

To handle the communication and measurement process among the ceiling beacons during the bootstrapping phase, and the distance measurement between beacons and the mobile entities during the service phase, we have developed a lightweight layered distance measurement stack, shown in figure 8.4. The bottom layer is an integrated layer that acts as a driver for the radio and the ultrasonic ranging hardware. The BMAC layer implements a simple CSMA medium access control protocol that provides reliable communication on the RF communication channel. The ranging coordination layer is implemented on top of BMAC and is responsible for the coordination of ultrasonic transmissions between nodes. This layer is responsible for ensuring that ultrasonic transmissions from different nodes do not interfere with each other. This is done by applying an arbitration protocol that determines the ordering that nodes transmit their reference signals. Finally, an additional layer above the ranging coordination layer performs an initial filtering on the measurement data and classifies the data into a set of appropriate data structures in which they are forwarded to the location computation engine, or in the case of a distributed implementation they are passed to the localization algorithm running in a separate task.

3. Location Discovery Algorithms

The two main location discovery algorithms devised for the Smart Kindergarten environment are *iterative multilateration* [4] and *collaborative multilateration* [6]. Both algorithms are primarily used for localizing the ceiling beacons during the bootstrapping phase. In iterative multilateration, a node localizes itself if it can receive reference signals from three other nodes that already know their locations. This allows the node with unknown location to estimate its position and then provide reference signaling for other nodes with unknown locations to estimate their locations. Given sufficient densities and an initial number of beacon nodes, iterative multilateration will localize all the nodes in the network. Iterative multilateration suffers from two main problems. First, it can get stuck in regions of the network that do not have sufficient beacon densities. If the process reaches a point where none of the nodes with unknown locations can obtain reference information from at least three beacon nodes, localization can get stuck. To alleviate this problem, we developed collaborative multilateration. Collaborative multilateration enables multiple nodes to share their beacon information and jointly estimate their positions using the beacon node locations and a set of inter-node distance measurements.

Collaborative multilateration uses a three-phase process. During the first phase, the nodes compute a set of initial estimates by forming a set of bounding boxes around the nodes. The nodes then organize themselves into over-constrained groups in which their positions are further refined using least squares. The refinement phase can use one of two possible computational models, centralized and distributed. The centralized computation model requires beacon positions and distance measurement information for the entire network. The distributed computation model is an approximation of the centralized computation model in which each node is responsible for computing its own location by exchanging information with its one-hop neighbors. The key attribute that makes distributed collaborative multilateration possible is the *in-sequence* execution within an over-constrained set of nodes. In distributed collaborative multilateration, each node executes a multilateration using the highly uncertain position estimates of its neighbors, and the corresponding distance to each of its neighbors. The consistent sequence of execution of multilaterations among each nodes results in the formation of a global gradient that allows each node to compute its own position estimate locally by following a gradient with respect to the global constraints.

3.1 Location Computation Engine

The location computation engine is the software the fuses distance measurements to derive node locations. This software is responsible for filtering the measurements and selecting the set of beacon nodes that will give the best position estimate based on a set of geometric criteria. Once the positions are computed they get stored in the central system database for further processing.

4. Conclusions and Lessons Learned

The design of the Smart Kindergarten localization infrastructure was a challenging and rewarding experience. Although the initial system achieves some level of functionality, this localization system is still under constant refinement. The design of customized wireless sensor node provided valuable insight in the design of low power platforms using off-the-shelf components. This process also revealed several sources of overhead in the low volume production of experimental systems in an academic environment. The purchasing of components from multiple vendors and unexpected lead-times in part delivery introduced unforeseen delays in the development our platforms. The experience derived from this project is driving the development of a new generation of localization platforms and algorithms. For more details and the latest developments in this platforms we refer the reader to our websites [1] and [8].

References

[1] ENALAB Website, http://www.eng.yale.edu/enalab

[2] S. Park, I. Locher, A. Savvides, M. B. Srivastava, A. Chen, R. Muntz, S. Yuen, *Design of a Wearable Sensor Badge for Smart Kindergarten*, Proceedings of the International Symposium on Wearable Computing, October 2002

[3] P. Steurer and M. B. Srivastava, *System Design of Smart Table*, Proceedings of IEEE International Conference on Pervasive Computing and Communications, PerCom 2003, March 2003.

[4] A. Savvides, C-C Han, and M. Srivastava. *Dynamic Fine-Grained Localization In Ad-Hoc Networks Of Sensors*. Proceedings of the Seventh Annual International Conference on Mobile Computing and Networking, July 2001. pp. 166 - 179

[5] A. Savvides and M. B. Srivastava, *A Distributed Computation Platform for Wireless Embedded Sensing*, Proceedings of International Conference on Computer Design (ICCD), Freiburg, Germany, September 2002

[6] A. Savvides, H. Park and M. B. Srivastava, *The n-hop Multilateration Primitive for Node Localization Problems*, Proceedings of Mobile Networks and Applications, 8, 443-451, 2003

[7] A. Savvides, W. Garber, R. L. Moses and M. B. Srivastava, *An Analysis of Error Inducing Parameters in Multihop Sensor Node Localization*, to appear in IEEE Transactions on Mobile Computing, 2004.

[8] Smart Kindergarten Website http://nesl.ee.ucla.edu/projects/smartkg

[9] M. Srivastava, R. Muntz and M. Potkonjak, *Smart Kindergarten: Sensor-based Wireless Networks for Smart Developmental Problem-solving Environments*, Proceedings of the ACM SIGMOBILE 7th Annual International Conference on Mobile Computing and Networking, Rome, Italy, July 2001

[10] A. Chen, R.R. Muntz, S. Yuen, I. Locher, S. Park. M.B. Srivastava, *A Support Infrastructure for the Smart Kindergarten*, IEEE Pervasive Computing, vol. 1, no. 2, April-June 2002, pp. 49-57

[11] Crossbow Website, http://www.xbow.com

Chapter 9

DIRECTIONS IN MULTI-QUERY OPTIMIZATION FOR SENSOR NETWORKS*

Alan Demers, Johannes Gehrke
Department of Computer Science
Cornell University, Ithaca, NY 14853
{ademers,johannes}@cs.cornell.edu

Rajmohan Rajaraman
College of Computer and Information Science
Northeastern University, Boston MA 02115
rraj@ccs.neu.edu

Niki Trigoni, and Yong Yao
Department of Computer Science
Cornell University, Ithaca, NY 14853
{niki,yao}@cs.cornell.edu

Abstract

The widespread dissemination of small-scale sensor nodes has sparked interest in a powerful new database abstraction for sensor networks: Clients "program" the sensors through queries in a high-level declarative language (such as a variant of SQL), and catalog management and query processing techniques abstract the user from the physical details of tasking the sensors. We call the resulting system a sensor data management system (SDMS). Sensor networks have important constraints on communication, computation and power consumption. Energy is the most valuable resource for unattended battery-powered nodes. Since radio communication consumes most of the available node power, our goal is to identify strategies that reduce network traffic. We give an overview of three distinct

*The authors wish to thank Douglas Holzhauer and Zen Pryke from the Air Force Rome Labs for helpful discussions. This work is supported by NSF Grants CCR-0205452, IIS-0133481, and IIS-0330201, by the Cornell Information Assurance Institute, and by Lockheed Martin.

approaches to reducing the cost of processing aggregate queries in sensor networks: i) selection of suitable routes for collecting results of multiple queries, ii) data reduction techniques that exploit query commonalities and iii) a hybrid pull-push communication paradigm for query and result propagation. We pay particular attention to the third approach and present in detail an algorithm for finding a pull-push configuration that minimizes on expectation the network traffic. Experimental analysis shows that our algorithm offers significant energy savings.

Keywords: Sensor networks, query processing, multi-query optimization.

1. Introduction

Sensor networks consisting of small sensor nodes with sensing, computation and communication capabilities are becoming ubiquitous. Large-scale, densely deployed sensor networks extend spatial coverage, accomplish higher resolution, and improve reliability. Sensor networks have many applications. For example, in the intelligent building of the future, sensors are deployed in offices and hallways to measure temperature, noise, and light, as part of the building control system. Another application is habitat monitoring. As an example, consider a biologist who may want to know of the presence of a specific species of birds, and once such a bird is detected, to map the bird's trail as accurately as possible. In this case, the sensor network is used for automatic object recognition and tracking.

Sensor networks have important constraints on communication, computation and energy consumption. First, the bandwidth of wireless links connecting sensor nodes is usually limited, on the order of a few hundred Kbps, and the wireless network that connects the sensors provides only limited quality of service, with variable latency and dropped packets. Second, sensor nodes have limited computing power and memory sizes that restrict the types of data processing algorithms that can be deployed. Third, wireless sensors have limited supply of energy, and thus power reduction is a major system design consideration.

Recently, a database approach to programming sensor networks has gained interest [1, 14, 21]: Clients "program" the sensors through queries in a high-level language (such as variants of SQL), and catalog management and query processing techniques abstract the user from the physical details of tasking the relevant sensors. There are several advantages to programming the sensor network through declarative queries. First, by letting the system decide and optimize the execution strategy of a query, the user is isolated from the physical properties of the network. Second, an important class of queries involves properties of *sets* of nodes, and thus declarative addressing of nodes in groups that satisfy user-defined properties is a natural way to interact with sensor data.

Third, the system can decide to perform in-network processing and optimize the resulting computation structure within the network by intelligently placing query operators. We call such a system a *sensor data management system*, or SDMS.

Most current SDMSs are optimized for executing a single aggregation query. Since radio communication consumes most of the available power of sensor nodes, SDMSs apply different strategies to minimize communication such as in-network processing. In practice, a SDMS supports many users accessing the sensor network simultaneously, with multiple queries running concurrently. The simple approach of processing every query independently of the others incurs redundant communication cost, draining the energy reserves of the nodes. We therefore need to investigate new approaches that take into consideration the query and sensor update workload in order to optimize data processing tasks in the SDMS.

In this chapter we overview several ongoing research directions in multiple query processing techniques for sensor networks. Our first direction is *aggregation tree selection.* Long-running aggregate queries are popular in sensor networks and have been discussed in recent papers [21, 14]. Processing aggregation queries requires that data from a set of sensor nodes be routed to the site where these queries were posed, with in-network aggregation on the route. The most natural way is along edges of a spanning tree that in some sense "embeds" all queries. Selection of suitable routes has an important effect on the cost of processing multiple queries.

A second, related research direction is *data reduction.* In order to minimize the number of messages for a given query workload and routing structure, we introduce a data reduction technique that exploits query commonalities at internal nodes where partial query results are merged. The feasibility and cost benefit of data reduction also depends in subtle ways on the choice of the routing tree.

Our third research direction is a *hybrid pull-push communication paradigm* for query and result propagation. There are two ways of executing queries. One possibility is to reactively send queries into the network and to *pull* relevant results out of the network. Another possibility is to proactively *push* all possibly relevant readings out of the network independent of the actually running set of queries. In this paper we propose a hybrid *pull-push* approach: sensor readings are pushed proactively to selected nodes in the network from which they are later pulled when queries are asked. Carefully drawing the line between the pull and the push areas can offer significant communication savings.

The remainder of the chapter is organized as follows: We first introduce our model of a sensor network in Section 2). We then overview three different approaches to energy-efficient processing of multiple aggregate queries in Section 3: these include aggregate tree selection, in-network data reduction, and hybrid pull-push communication. In Section 4, we focus on the third approach and present in detail an algorithm that selects an energy-optimal pull-push configuration given a query and sensor update workload. Section 5 surveys related work and Section 6 concludes the paper and discusses ideas for future work.

2. Model

In this section, we describe our model for sensor networks and sensor data, and outline our architectural assumptions. **Sensor Networks.** We consider a sensor network that consists of a large number of *sensor nodes* connected through a multi-hop wireless network [16, 10]. We assume that nodes are stationary, all node radios have the same fixed communication range, and that each node is aware of its own location. (Note that future generations of nodes might have variable-range radios; an extension of this work to variable-range radios is future work.) We distinguish a special type of node called a *gateway node*. Gateway nodes are connected to components outside of the sensor network through long-range communication (such as cables or satellite links), and all communication with users of the sensor network goes through the gateway node. Sensor networks have the following physical resource constraints:

Communication. The bandwidth of wireless links connecting sensor nodes is usually limited, on the order of a few hundred Kbps; the network provides limited quality of service, with variable latency and large packet drop probability.

Power consumption. Sensor nodes have limited supply of energy; thus, energy-efficiency is a major design consideration.

Computation. Sensor nodes have limited computing power and memory sizes that restrict the types of data processing algorithms that can be deployed and intermediate results that can be stored on the sensor nodes.

Sensor Data. Each sensor can be viewed as a separate data source that generates structured records with several fields such as the id and location of the sensor that generated the reading, a time stamp, the sensor type, and the value of the reading. (We assume that some of the signals might have been postprocessed by a signal processing layer.) Conceptually, we view the data distributed throughout the sensor network as forming a distributed database system consisting of multiple tables with different types of sensor data.

Queries. The sensor network is programmed through declarative queries which abstract the functionality of a large class of applications into a common interface of expressive queries. Our work does not depend on any specific query language; instead it applies to any query processing strategy that performs in-network processing by collecting data from multiple sensors onto a designated subset of nodes. These nodes may simply store unprocessed sensor readings directly, or they may materialize the result of more complex processing over the sensor readings.

Aggregation queries are particularly important in energy-constrained sensor networks since they involve propagation of data summaries, rather than of individual sensor readings. Since many applications are interested in monitoring an environment over a longer time-period, *long-running queries* that periodically produce answers about the state of the network are especially important. Users might also be interested in issuing *snapshot queries* that require an answer only about the current state of the physical environment.

Synchronization Between Sensors. We assume that the clocks of neighboring nodes in the sensor network are reasonably synchronized, either through GPS or through distributed time synchronization algorithms (e.g., [12, 4]).

3. Multi-Query Optimization

In this section, we give an overview of three distinct approaches to energy-efficient query processing in sensor networks: tree selection, data reduction and hybrid push-pull data dissemination. We first define some useful notation and then give intuitive examples that illustrate the optimization opportunities arising in each of the approaches.

We consider a set of sensor nodes $s_1, \ldots, s_k$ spread in the plane. We divide time into *rounds*. Queries are executed at the end of a round, and they refer to sensor readings generated during that round. Let $Q = \{(q_1, p_1), \ldots, (q_n, p_n)\}$ be the query workload, where p_i is the probability that q_i is executed at the end of a round. Similarly, let $S = \{(s_1, u_1), \ldots, (s_k, u_k)\}$ represent the sensor update workload, where u_i is the probability that sensor s_i is updated during a round. For concreteness, we consider queries q_i of the form:

$$select\ aggr(s.attr)\ from\ Sensors\ s\ where\ s.loc\ in\ Region_i.$$

Each query returns the aggregate value of some subset of the data sources lying in a certain region. We assume that the aggregate function used is the same for all queries and we specifically consider the sum function in the examples that follow.

We are given a dissemination tree of sensors rooted at the special node called the gateway. The energy cost of sending an n-bit message along a tree edge is $\alpha + \beta n$, where α is the startup cost of activating an edge and β represents the per-bit cost. Both the parameters α and β are sums of two components each, $\alpha_s + \alpha_r$ and $\beta_r + \beta_s$, respectively, where α_s (respectively, β_s) is the part of the fixed cost (respectively, per-bit cost) associated with the sender and α_r (respectively, β_r) is the part associated with the receiver.

Most SDMSs are optimized for executing a single long-running query, which is processed in two phases: In the *query propagation phase*, the query is disseminated from the gateway node to the relevant sensor nodes. This phase results in the construction of a dissemination tree, e.g. by letting every sensor select as its parent the node on the shortest path to the gateway. In the *query execution phase*, results are periodically pushed from the sensor nodes to the gateway node along the tree paths, performing in-network aggregation on the way. In the remainder of this section, we show that significant energy savings can be achieved by applying multi-query optimization techniques that take into consideration the query and sensor update workloads.

3.1 Tree Selection

The effectiveness of processing multiple aggregate queries along a dissemination (or aggregation) tree depends on the choice of the aggregation tree. In this subsection, we study *tree selection*, the problem of computing an optimal tree in the underlying sensor network that connects the root with the set of active sensors.

We consider two performance criteria for determining the quality of an aggregation tree, both based on energy consumption. One criterion is to minimize the *total energy consumed* for processing the given query set. Another criterion is to minimize the *maximum energy consumed at a node*, aimed at maximizing the network lifetime (defined as the time until the first node dies). In the following, we make the simplifying (if somewhat unrealistic) assumption that the start-up cost of activating an edge (α) is 0; that is, the energy consumed in the transmission and reception of b bits on a link is proportional to b, thus implying that minimizing the total number of data bits sent is equivalent to minimizing the total energy consumed.

We first note that the general problem of tree selection in such a network for minimizing total energy consumed is intractable: an easy reduction from the NP-complete rectilinear Steiner tree problem [5] shows that tree selection is NP-hard even for the case of a single query. In the remainder of this section, we present a series of examples that indicate how the effectiveness of aggregation trees varies with problem instances and the performance criteria. In our examples, we assume that nodes of the sensor network are organized in a grid.

A given instance specifies an underlying sensor network G with a designated root r and a set Q of queries covering a set S of sensors. For a given tree T that connects every sensor node included in a query in Q to the root, the total cost (or the maximum cost) depends on the query processing algorithm used over T, since different solutions may differ in the amount of communication along the tree links. Here we assume that the total cost (respectively, maximum cost) is that of an optimal query processing plan for the set Q.

Example: Consider a sensor network consisting of a 3×3 grid with 9 sensor nodes. We let (i, j) denote the node in row i, column j, $0 \leq i, j < 3$. Suppose the root server is the center $(1, 1)$. We consider different query set instances, in each of which all the sensor nodes are active; thus the desired tree is, in fact, a spanning tree. We assume that all the update probabilities and the query probabilities are 1.

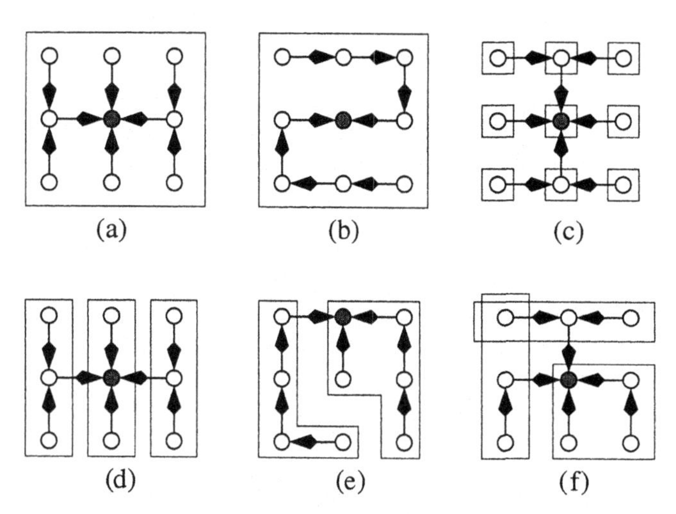

Figure 9.1. Aggregation trees and query set instances. Each of the six figures illustrates a query set instance and an aggregation tree. In each figure, a query is represented by the set of nodes within a bounding box and the aggregation tree is shown with edges directed toward the root, which is shaded gray. For reference, we label the query sets and trees in parts (a) through (f) by Q_a through Q_f and T_a through T_f, respectively.

1 Q_a contains the single set of all nodes. That is, the only query of interest is an aggregate of all the nodes. In this case, every spanning tree has the same total cost and is hence optimal. In particular, the tree T_a consisting of all the vertical edges together with the horizontal edges on row 1 is an optimal tree (see Figure 9.1(a)).

2 Q_b is the same as Q_a, but the performance criterion is is network lifetime rather than total cost. Tree T_a is no longer optimal. In a "snakelike" tree T_b (see Figure 9.1(b)), each node sends and receives at most one message and the root receives two messages, while in T_a, certain intermediate nodes receive two messages and send one message and the root receives four messages. The exact benefit of T_b over T_a depends on the ratio of β_s to β_r.

3 Q_c contains all the singleton sets. That is, every query seeks the data on a single node. In this case, a tree with optimal total cost is a shortest path tree. Thus, tree T_a has optimal total cost. The same can be said for tree

T_c consisting of all the horizontal edges together with the vertical edges on column 1 (see Figure 9.1(c)). Both the trees have a cost equal to the sum of distances from each node to the center. In the given network, it is 12; for a general $N \times N$ grid the cost is $\Theta(N^3)$. Clearly, the snakelike tree T_b is very poor, incurring a total cost of $\Theta(N^4)$.

4 Q_d equals $\{\{(i, j) : 0 \leq i < 3\} : 0 \leq j < 3\}$. That is, every query is an aggregate on a column. In this case, one can see that tree T_d (same as T_a) is a tree with optimal total cost 8 (see Figure 9.1(d)) while tree T_c has total cost 12 and is suboptimal.

5 In all the above instances, a tree with optimal total cost is a shortest path tree. However, this is not necessary. Suppose the root is at node $(0, 1)$ and the query set Q_e consists of the query $\{(0,0), (1,0), (2,0), (2,1)\}$ and a set consisting of each of the other nodes (see Figure 9.1(e)). For this instance, in a tree T_e with optimal total cost, the node $(2, 1)$ is connected to the root $(0, 1)$ via the long path $(2, 1) \rightarrow (2, 0) \rightarrow (1, 0) \rightarrow (0, 0) \rightarrow (0, 1)$.

6 As a final example, we consider a query set Q_f with root $(1, 1)$ in which the queries are not disjoint (see Figure 9.1(f)). For this instance, the tree T_a has total cost 10. This is because the bases of the projection of the queries on to columns 0 and 2 are both of size two; consequently, the two horizontal edges on row 1 have to carry 2 information units, one corresponding to a column and the other corresponding to a node in row 0. In contrast, tree T_f has a cost of 9 only because only one subtree rooted at a child of the root has a basis of size more than one; in particular, only the vertical edge $((0, 1), (1, 1))$ carries two units of information, while every other edge carries one.

The above examples indicate that the optimal trees have diverse characteristics, depending on the particular query instance and the performance criterion being considered, even for the special case when the set of active sensors includes all the nodes.

To close this subsection, we note that it is not necessary to confine our attention to trees. When queries are not disjoint, the generalization to DAGs can be beneficial. It is not difficult to generalize the example of T_f above so that a DAG in which the shared sensor value is sent to two neighbors yields a lower cost solution than any tree. Aggregation DAGs are the subject of future research.

3.2 Data reduction

In the previous subsection, we showed that the choice of dissemination tree has an important effect on the cost of query processing. Our implicit assumption was that given a certain tree, the optimal multi-query processing plan is selected. In this subsection, we elaborate further on this assumption, i.e. we study the problem of identifying an optimal plan for processing multiple aggregate queries along a given dissemination tree.

Figure 9.2 shows a simple example of a dissemination tree that connects five sensor nodes. Here the root r is the gateway node, the leaves a, b and c are the data sources, and values must be routed from the data sources through internal node i to the root as needed to compute query results. Data sources are located at the leaf nodes only for the purposes of this example; in general, we are able to deal with any intermediate node generating sensor readings.

In the remainder of this section we present two scenarios of query evaluation on the tree of Figure 9.2. Each scenario introduces one of our techniques for multi-query optimization.

Deterministic Updates: For the simplest scenario, suppose that each sensor produces a new value in each round. We call this the "D" scenario: it assumes that data updates occur deterministically, with probability 1. In each round, each sensor node proactively sends its current sensor value to its parent in the communication tree. Interior nodes of the tree compute sub-aggregates of the values they receive from their children, and forward them up the tree towards the root. Under these conditions, multi-query optimization involves recognizing when the values of sub-aggregates can be shared effectively among queries, so that redundant data messages can be eliminated.

Consider evaluating the three queries $a + b$, $a + b + c$, and c on the tree of Figure 9.2. In each round all leaves send their values to the interior node i, which then has enough information to compute the values of all the queries. However, given our assumption that the energy cost of local computation is negligible compared to the cost of communication, it would be wasteful to forward all these values from i to the root. The three queries $a + b$, $a + b + c$, and c are not linearly independent – the values of any two of them can be used to calculate the value of the third. Thus, node i should forward only *two* of the values (say $a+b$ and c). The root should then recompute the third value $(a+b+c)$ locally, achieving a net saving of energy.

In general, this technique of "reducing" the set of data values forwarded toward the root can be repeated at every subtree. Queries in the workload are first projected to a subtree rooted at node N which contains, say, ℓ sensor descendants. The projection is represented as a $n \times \ell$ bit matrix M, where $M(i, j)$ is 1 if query Q_i accesses sensor s_j. M is reduced to echelon form. The row cardinality (rank) of the reduced matrix M' denotes the number of results that node N must send to its parent. This observation leads to an algorithm that minimizes communication cost.

Irregular Updates: Next we consider the "I" scenario, in which sensor updates occur occasionally according to some probability distribution. The goal of the query optimizer is to choose an efficient "result encoding," sending the minimum amount of data up the tree that will enable the root both to identify the queries affected by the updated sensors and to compute the values of those queries. We call data sent for these purposes RESULTCODE and RESULT-DATA, respectively.

Returning to the example in Figure 9.2, consider the same three queries discussed for the DD scenario: $a + b$, $a + b + c$, and c. Suppose only sensor a is updated in a round. Clearly the information forwarded up edge (i, r) must inform the root that queries $a + b$ and $a + b + c$ are affected, and must include the current value of a. However, this does not imply that the root must "know" the exact set of sensors that were updated. It is easy to verify that sensors a and b and the aggregate $a + b$ are all indistinguishable by any of the queries in the workload. Consequently, rounds in which a changes, or b changes, or both a and b change, can all result in identical messages being sent along edge (i, r).

Thus, in the I scenario the goal of the multi-query optimizer is to find an optimally compressed result encoding that eliminates unnecessary distinctions in the RESULTCODE component in addition to representing the RESULTDATA component efficiently. This is a more difficult problem than in the D scenario, since the expected performance of a result encoding clearly depends on the distribution of sensor updates.

3.3 Hybrid pull-push model

The norm in existing sensor networks is either to proactively push sensor readings towards the gateway (*push* model) or to propagate currently running queries to the network and pull relevant data on demand (*pull* model). The question arises whether there are workloads that would favor a hybrid *pull-push* approach. In a hybrid model, sensor readings are pushed proactively into the network and stored at carefully selected nodes, from which they are later pulled when queries are asked. By carefully selecting the borderline that separates the pull from the push areas of the network we can achieve significant communication savings when queries and sensor updates are probabilistic.

Let tree edges be classified either as *proactive* (push) or *on-demand* (pull) in the optimal tree configuration. Along a proactive edge, all new data is sent upward unconditionally in every period. Thus, there is no need to send request messages down proactive edges. In contrast, an on-demand edge transfers only the data required to answer the queries posed in the current round. Thus, an on-demand edge requires an explicit request message in each period. The request message must be sent even if no partial result is required in the current period. This is a consequence of a (realistic) energy model in which radio receivers have substantial power requirements. In the presence of collisions at the MAC layer and imperfect clock synchronization, the energy cost (at the listener) of determining that no message will arrive in a period can be substantially more than the energy cost (at sender and listener) of transferring a short "nothing to request" message.

For a proactive edge, the per-period expected cost is determined by the probability of new sensor readings being generated in the subtree beneath it. For an on-demand edge, the expected cost is the cost of its (unconditional) request message plus the cost of partial result messages needed to answer the currently posed queries.

Let $Q = \{(a{+}b, 1{-}\epsilon), (a{+}c, 1{-}\epsilon)\}, S = \{(a, 1), (b, 1), (c, 1), (i, 1), (r, 1)\}$ be the traffic workload in the tree of figure 9.2. We show that the optimal pull-push configuration depends on the expected query and result costs at each edge. As in our earlier discussion of tree selection, we assume a simple communication energy model in which the edge activation cost (α) is 0 and the per-bit cost (β) is 1. Let q be the expected cost of sending a query request message down a tree edge, and $r >= q$ the cost of sending a data result message. The expected pull cost of the edge between the root and interior node i is $q + 2(1 - \epsilon)r$, representing an unconditional request message and two query results with independent probabilities $(1 - \epsilon)$. The expected push cost of this edge is $2r$. The edges entering nodes b and c have equal expected pull costs of $q + (1 - \epsilon)r$. The edge entering leaf a has expected pull cost $q + (1 - \epsilon^2)r$. The expected push cost of edges near the leaves is r.

The optimal hybrid solution is now completely determined by the relative values of q, r and ϵ. When $q > 2\epsilon r$, a completely proactive (push) solution is best. For $\epsilon r < q < 2\epsilon r$, the best solution is to make on-demand (pull) only the edge from the root to node i, proactively sending data from the leaves to node i. For $\epsilon^2 r < q < \epsilon r$ it becomes beneficial to make the edges to nodes b and c on-demand as well, but still materialize the value of a at node i. And for $q < \epsilon^2 r$ (the second limiting case above),a completely on-demand solution is optimal. This example illustrates that query probabilities affect our choice of where to draw the line between the proactive and on-demand edges. Similar examples can be given to show the effect of sensor update probabilities.

Pull-push decisions were made based on the expected costs of query and result messages at different edges. In the general case, multi-query optimization techniques complicate the task of computing these costs. Recall the query and sensor update workloads in the example of the previous subsection: $Q = \{(a + b, 1), (a + b + c, 1), (c, 1)\}$ and $W = \{(a, 1), (b, 1), (c, 1), (i, 1), (r, 1)\}$. Since queries are deterministic a push strategy is preferable: each of the nodes a, b and c must push one value to its parent i, and i needs to push two values (instead of three) to the root r. Under deterministic query and sensor update workloads the number of results that each node needs to forward to its parent is easy to evaluate (as the rank of the corresponding projected query matrix). However, when queries are probabilistic, the expected rank of the projected query matrix is very hard to compute. Things only become more complex when sensor updates are probabilistic. Here, *current* queries are projected only to the *updated* sensors of a subtree, and the corresponding bit matrix is reduced. Enumerating all different combinations of queries and sensors that might occur in a round and evaluating the rank of the corresponding matrices is prohibitive, especially for sensor nodes with limited processing capabilities. The expected traffic routed over an edge is critical, however, in deciding whether to pull or push data along the edge.

4. An adaptive hybrid pull-push approach

In the previous section, we discussed three different approaches to saving energy in processing multiple aggregate queries. In this section, we focus on the third approach: we present in detail an algorithm that selects an optimal pull-push configuration, discuss its complexity and give a set of preliminary experimental results.

4.1 Algorithm

Our algorithm works in two phases and selects an optimal pull-push strategy that addresses two issues: i) depending on the aggregate function and the multi-query optimization techniques applied, the expected volume of local edge traffic could be hard to compute; ii) by synthesizing locally-optimal decisions based on the expected traffic, we might obtain an incompatible pull-push configuration (e.g. select a pull edge below a push edge).

Simulation phase: This is a statistics gathering phase. Nodes monitor the traffic of the network for a certain number of rounds (say m rounds). At every round, each node keeps record of the query q and result traffic r routed through it using the pull model. It also calculates the size of results R that it would forward to its parent, had it *not* known the current queries, i.e. had it used the push model. At the end of m rounds, every node evaluates the local average sizes $avg(q)$, $avg(r)$ and $avg(R)$ of the forwarded query and result messages.

Dynamic Programming (DP) phase: By the end of the simulation phase, every node N has evaluated the average cost of applying the push or pull model at the local edge $e_{N \leftrightarrow P(N)}$: $Push_N = \alpha + \beta * avg(R)$ and $Pull_N = 2 \times \alpha + \beta * (avg(q) + avg(r))$. The optimal compatible pull-push configuration is selected in two passes. In the bottom-up pass, every node N recursively evaluates and informs its parent about the following costs:

$STPull_N = Pull_N + \sum_{i=1}^{childNo_N} STPullPush_{ch_N[i]}$

$STPush_N = Push_N + \sum_{i=1}^{childNo_N} STPush_{ch_N[i]}$

$STPullPush_N = min\{STPull_N, STPush_N\}$

In the top-down pass, a node N (except for the root) waits until it is informed about the *model* (pull or push) used by its parent. Initially *model=pull* for every node. The *model* value of the current node is set to *push* if either $model_{P(N)} = push$ or $STPush_N < STPull_N$. Node N then broadcasts its own *model* value to its children. By the end of the top-down phase, an optimal pull-push model has been assigned.

Algorithm evaluation: We use the standard technique of Monte Carlo simulations to obtain near-accurate estimates of the required query and result costs. Let $Y(U_1, \ldots, U_k)$ be a function of independent query or update events U_i. Y may stand for either q, r or R cost. Let $\overline{Y}$ denote $\sum_j Y_j/m$ of m samples from the underlying probability space. If $Y_{\max}$ is the maximum possible value for Y, then for any $0 < \varepsilon < 1$, $\Pr[|\overline{Y} - E[Y]| > \varepsilon] \leq e^{-\varepsilon^2 m/Y_{\max}}$. The proof is omitted for lack of space. By setting the number of samples m sufficiently larger than $Y_{\max}$, we can set the probability that the estimate is ε-away from the expectation arbitrarily close to zero. The independence assumption for query and update probabilities is used to bound the error between the estimated and the expected values. The proposed adaptive algorithm would be applicable even without the assumption, but without providing optimality guarantees. The DP phase enforces compatibility constraints for a hybrid model, i.e. an edge can be assigned the pull model, iff all ancestor edges are also pull. It can be implemented in a distributed manner and communication-wise it involves sending two small messages per node. The calculations that it involves are very simple and do not require large storage capabilities. In order to adjust to changes in traffic probability distributions, the adaptive algorithm is repeated periodically.

4.2 Experiment

We simulate a network of 400 nodes organized in rectangular grid. A tree connects all nodes to the gateway (located in the bottom left corner). In every round, we run multiple sum queries that cover all sensors in a rectangular area. The area dimensions are randomly chosen between 1 and 20. Query messages are bit-vectors denoting which queries in the probabilistic workload QW occurred at the current round. Result messages include a bit-vector, which denotes which sensors in the subtree are updated in the current round, and a set of reduced query results. Each query result has size 32 bits.

We first study the performance of different models on a workload of 50 queries with small probabilities (0.1). Figure 9.3 shows that the push method outperforms the pull method for low update probabilities. The hybrid pull-push method outperforms the others in all cases, offering benefits of up to 20%. Figure 9.4 shows the impact of query probabilities when the update probability is low (0.1). For query probability close to 0.2, the pull and the push cost become equal and the relative benefit of the hybrid approach (25%) is maximized. We finally evaluate the role of the edge activation cost α for small query probabilities (0.1) and deterministic updates (figure 9.5). When

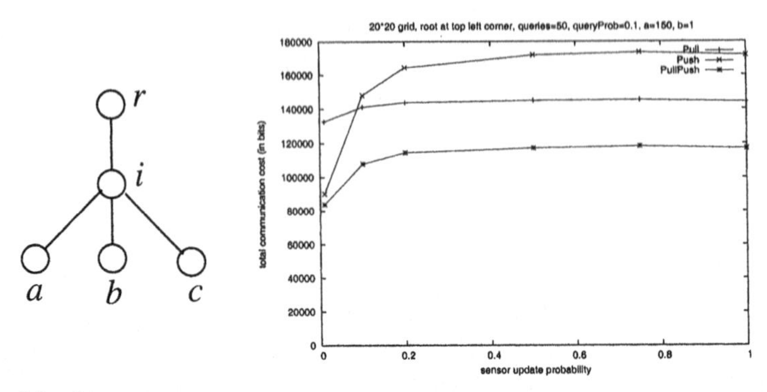

Figure 9.2. An tree example.

Figure 9.3. Impact of update probabilities

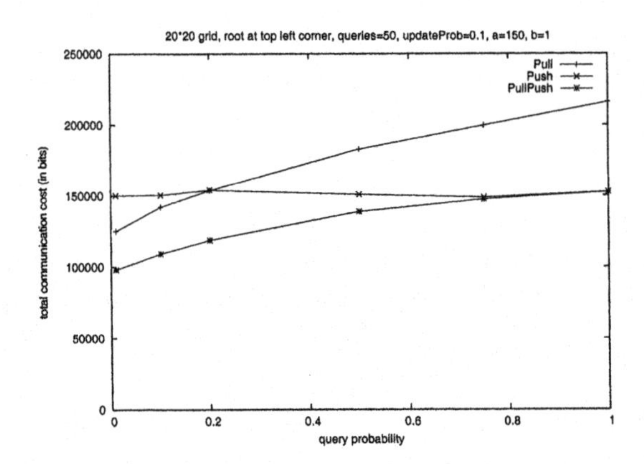

Figure 9.4. Impact of query probabilities

α is very small pull outperforms push, since the overhead of query requests is small compared to their filtering benefits. Push is preferred however for large αs. The relative benefit of the hybrid method is maximized when the pure method costs are equal ($\alpha = 210$). This point shifts to the left, if we increase query or decrease update probabilities.

5. Related Work

There is much existing work related to query processing in sensor networks, both in database and network communities. We discuss some of this work below.

Several research groups have focused on in-network query processing as a means of reducing energy consumption. The work most closely related to ours is the TnyDB project [14] at U.C. Berkeley, which investigates query process-

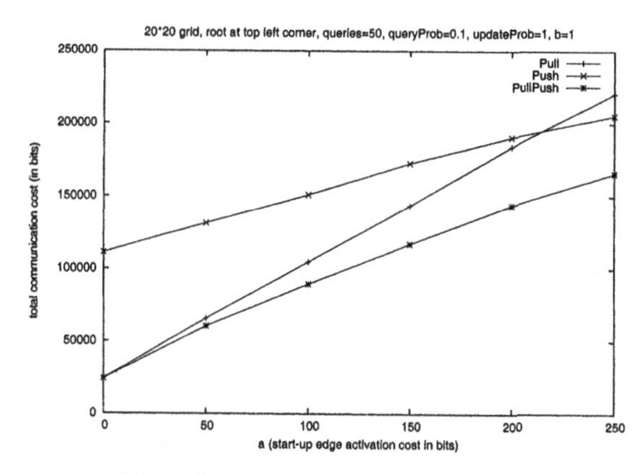

Figure 9.5. Impact of edge start-up cost

ing techniques for sensor networks including an implementation of the system on the Berkeley motes and aggregation queries An acquisitional approach to query processing is proposed in [13] for query optimization, in which the frequency and timing of data sampling is discussed. The sensor network project at USC/ISI group [10, 8, 7] proposes an energy-efficient aggregation tree using data-centric reinforcement strategies (directed diffusion). A two-tier approach (TTDD) for data dissemination to multiple mobile sinks is discussed in [22]. In a recent study [6], an approximation algorithm has been designed for finding an aggregation tree that simultaneously applies to a large class of aggregation functions. Our study differs from previous work in that we consider multi-query optimization for sensor networks, which in conjunction with single query optimization techniques significantly reduces energy consumption and improves network lifetime.

Although there has been much work on query processing in distributed database systems, [24, 2, 15, 23, 11], there are major differences between sensor networks and traditional distributed database systems. Most related is work on distributed aggregation, but existing approaches do not consider the physical limitations of sensor networks [17, 20].

The data dissemination algorithms studied in this chapter are all aimed at minimizing energy consumption, a primary objective in communication protocols designed for sensor (and ad hoc) networks. Recent work on energy-aware routing proposes the selection of routes on the basis of available energy in order to increase network lifetime [3, 25, 16]. Heinzelman et al. present the SPIN family of network protocols for communication of large messages in sensor networks [9]. GAF (Geographical Adaptive Fidelity) [19, 18] is an algorithm that also conserves energy by identifying nodes that are equivalent from a routing perspective and then turning off unnecessary nodes. While the preceding studies consider routing protocols for arbitrary communication patterns, our study focuses on minimizing the communication cost associated with processing multiple aggregate queries over a given communication tree. Although our techniques are mainly designed for tree-based routing structures, they can be integrated into other communication protocols to reduce the communication cost due to multiple concurrent queries.

6. Conclusion and Future Work

In this chapter, we identified three different approaches to optimizing the execution of aggregate queries in sensor networks. Their common objective is to reduce the network traffic, and therefore the energy consumption at the nodes. We first showed that careful selection of the aggregation tree that connects the sensor nodes to the gateway plays an important role in reducing communication cost. For a given aggregation tree, further savings can be achieved by reduction techniques that exploit commonalities among queries and consider irregular sensor updates. Finally, we proposed a hybrid pull-push paradigm for reducing data dissemination cost relative to traditional pull or push techniques. Our detailed description of the third approach shows that it can easily be implemented in a distributed manner and given the limited memory and computational capabilities of small sensor nodes. The effectiveness of our technique was illustrated by some initial experimental results.

In the future we plan to explore further these promising directions and devise approximation algorithms for selecting an aggregate tree and reducing data disseminated along its paths. We believe that important progress can be made if we exploit not only the query correlations, but also the correlations of the physical sensed data. Moreover, the uncertainty involved in sensing physical phenomena gives some degree of freedom in delivering approximate query answers, which in turn, could lead to energy savings in large sensor deployments. Finally, we would like to consider multi-query optimization for a larger class of queries occurring in tracking, emergency and other sensor network applications.

Acknowledgments

The authors wish to thank Douglas Holzhauer and Zen Pryke from the Air Force Rome Labs for helpful discussions. This work is supported by NSF Grants CCR-0205452, IIS-0133481, and IIS-0330201, by the Cornell Information Assurance Institute, and by Lockheed Martin.

References

[1] Bonnet, P., Gehrke, J., and Seshadri, P. Querying the physical world. *IEEE Personal Communications*, 7(5):10–15, 2000.

[2] Ceri, S. and Pelagatti, G. *Distributed Database Design: Principles and Systems.* MacGraw-Hill (New York NY), 1984.

[3] Chang, J. and Tassiulas, L. Energy conserving routing in wireless ad-hoc networks. In *Proceedings of the 2000 IEEE Computer and Communications Societies Conference on Computer Communications (INFOCOM-00)*, pages 22–31, Los Alamitos. IEEE, 2000.

[4] Elson, J., Girod, L., and Estrin, D. Fine-grained network time synchronization using reference broadcasts. In *Proceedings of the 5th Symposium on Operating Systems Design and Implementation*, pages 147–163, 2000.

[5] Garey, M. and Johnson, D. The rectilinear Steiner tree problem is NP-complete. *SIAM Journal on Applied Mathematics*, 32:826–834, 1977.

[6] Goel, A. and Estrin, D. Simultaneous optimization for concave costs: Single sink aggregation or single source buy-at-bulk. In *Proceedings of the 14th Annual ACM-SIAM Symposium on Discrete Algorithms*, 2003.

[7] Heidemann, J., Silva, F., Yu, Y., Estrin, D., and Haldar, P. Diffusion filters as a flexible architecture for event notification in wireless sensor networks. Technical Report ISI-TR-556, USC/Information Sciences Institute, 2002.

[8] Heidemann, J.S., Silva, F., Intanagonwiwat, C., Govindan, R., Estrin, D., and Ganesan, D. Building efficient wireless sensor networks with low-level naming. In *Symposium on Operating Systems Principles*, pages 146–159, 2001.

[9] Heinzelman, W.R., Kulik, J., and Balakrishnan, H. Adaptive protocols for information dissemination in wireless sensor networks. pages 174–185. ACM SIGMOBILE, ACM Press, 1999.

[10] Intanagonwiwat, C., Govindan, R., and Estrin, D. Directed diffusion: A scalable and robust communication paradigm for sensor networks. pages 56–67. ACM SIGMOBILE, ACM Press, 2000.

[11] Kossmann, D. The state of the art in distributed query processing. *Computing Surveys*, 32, 2000.

[12] Liao, C., Martonosi, M., and Clark, D. Experience with an adaptive globally-synchronizing clock algorithm. In *Proceedings of the 11th Annual ACM Symposium on Parallel Algorithms and Architectures*, pages 106–114, 1999.

[13] Madden, S., Franklin, M.J., Hellerstein, J., and Hong, W. The design of an acquisitional query processor for sensor networks. In *Proceedings of the ACM SIGMOD International Conference on Management of Data*, 2003.

[14] Madden, S.R., Franklin, M.J., Hellerstein, J.M., and Hong, W. Tag: A tiny aggregation service for ad-hoc sensor networks. In *OSDI*, 2002.

[15] Özsu, M.T. and Valduriez, P. *Principles of Distributed Database Systems.* Prentice Hall, Englewood Cliffs, 1991.

[16] Pottie, G.J. and Kaiser, W.J. Embedding the Internet: wireless integrated network sensors. *Communications of the ACM*, 43(5):51–51, 2000.

[17] Shatdal, A. and Naughton, J. Adaptive parallel aggregation algorithms. In Carey, Michael J. and Schneider, Donovan A., editors, *Proceedings of the 1995 ACM SIGMOD International Conference on Management of Data*, pages 104–114, San Jose, California, 1995.

[18] Xu, Y., Bien, S., Mori, Y., Heidemann, J., and Estrin, D. Topology control protocols to conserve energy inwireless ad hoc networks. Technical Report 6, University of California, Los Angeles, Center for Embedded Networked Computing. submitted for publication, 2003.

[19] Xu, Y., Heidemann, J., and Estrin, D. Geography-informed energy conservation for ad hoc routing. In *Proceedings of the ACM/IEEE International Conference on Mobile Computing and Networking*, pages 70–84, 2001.

[20] Yan, W.P. and Larson, P. Eager aggregation and lazy aggregation. In Dayal, Umeshwar, Gray, Peter M. D., and Nishio, Shojiro, editors, *VLDB'95, Proceedings of 21th International Conference on Very Large Data Bases*, pages 345–357, Zurich, Switzerland. Morgan Kaufmann, 1995.

[21] Yao, Y. and Gehrke, J. Query processing in sensor networks. In *Proceedings of the the First Biennial Conference on Innovative Data Systems Research (CIDR 2003)*.

[22] Ye, F.L., Haiyun, C., Jerry, L.S., and Zhang, L. A two-tier data dissemination model for large-scale wireless sensor networks. In *Proceedings of the Eighth Annual International Conference on Mobile Computing and Networking (MobiCom)*, 2002.

[23] Yu, C. and Meng, W. *Principles of Database Query Processing for Advanced Applications*. Morgan Kaufmann, San Francisco,1998.

[24] Yu, C. T. and Chang, C. C. Distributed query processing. *ACM Computing Surveys*, 16(4):399–433, 1984.

[25] Yu, Y., Govindan, R., and Estrin, D. Geographical and energy aware routing: A recursive data dissemination protocol for wireless sensor networks. Technical Report UCLA/CSD-TR-01-0023, University of Southern California, 2001.

Chapter 10

UBIQUITOUS VIDEO STREAMING: A SYSTEM PERSPECTIVE

Mario Gerla, Ling-Jyh Chen, Tony Sun, and Guang Yang
Department of Computer Science
University of California at Los Angeles, Los Angeles, CA 90095
{gerla,cclljj,tonysun,yangg}@cs.ucla.edu

Abstract Video streaming has become a popular form of transferring video over the Internet. With the emergence of mobile computing needs, a successful video streaming solution demands 1) uninterrupted services even with the presence of mobility and 2) adaptive video delivery according to current link properties. In this paper we study the need and evaluate the performance of adaptive video streaming in vertical handoff scenarios. We created a simple handoff environment with Universal Seamless Handoff Architecture (USHA), and used Video Transfer Protocol (VTP) to adapt video streaming rates according to the "Eligible Rate Estimates". Using testbed measurements experiments, we verify the importance of service adaptation, as well as show the improvement of user-perceived video quality, via adapting video streaming in the vertical handoffs.

Keywords: Adaptive video streaming, VTP, seamless handoff, vertical handoff.

1. Introduction

As the demand, production and consumption of digitized multimedia has intensified in recent years, the latest application trends have created an increasing interest in providing practical multimedia streaming systems to meet the needs of mobile computing. In order to provide uninterrupted services and maximum user-perceived quality, a successful video streaming solution needs to adapt appropriately to mobile handoff scenarios.

Consider the scenario where a user is in the midst of monitoring her biological experiment through a multimedia streaming broadcast, on a PDA device, via an 802.11b wireless connection in her office. Concurrently, she is informed of an urgent request for her immediate presence from her collaborators 20 miles away. Whereas she cannot afford to miss neither the streaming multimedia nor her meeting with her collaborators, an ideal ubiquitous computing solution would allow her to continue her current multimedia streaming session while in transit from her current location to her final destination. This involves leaving her office with her PDA (departing from an existing 802.11b high-capacity connection), take an express shuttle to her collaborator's location (during which time continuing her monitoring via a lower-capacity 1xRTT connection with the multimedia quality adapted to the changed capacity), and arrive at her collaborator's office (entering another 802.11b network and receiving multimedia of higher quality again). Although visions of such system have existed for some time [7] [20] however, an actual implemented system capable of handling the above scenario was not previous explored.

As previously identified by [7] [20], in order to provide a system that addresses quality of service in mobile computing environments, the following key issues need to be resolved: 1) seamless mobility across heterogeneous networks, 2) application adaptation to maximize the end user's perceived quality, and 3) adaptation to network dynamics such as wireless channel errors and congestion.

To accommodate mobile users switching between networks of different capacities, a seamless handoff technology, that preserves existing application sessions, is needed to tackle the first issue. Since mobile users may roam in an arbitrary pattern, an adaptive multimedia streaming technology, capable of maximizing the end user's perceived quality, is needed to address the second and third issues. Combining the criterion discussed above, a complete ubiquitous video streaming solution will undoubtedly incorporate both seamless handoff and adaptive multimedia streaming technologies.

For the purpose of this system, a simple seamless handoff environment is created with Universal Seamless Handoff Architecture (USHA)[6] to handle various handoff scenarios. An important feature of USHA is application session persistence. USHA can quickly adapt to user mobility while maintaining uninterrupted connectivity for established network sessions. Furthermore, USHA requires little modification to the current Internet Infrastructure, making it an attractive choice for a seamless handoff testbed. We will discuss USHA in more details in section 3.

A video streaming protocol, Video Transport Protocol (VTP) [2] is used for adaptive streaming applications. VTP adapts its sending rate, and thus quality, according to network conditions. Generally speaking, video streams encoded at higher rates have better quality over those encoded at lower rates, but they also demand more bandwidth. On the Internet where cross traffic is highly dynamic, bandwidth may not always be able to meet the demand. In such cases, the streaming server must lower its sending rate, or its packets would be heavily lost, severely impairing the quality perceived by the end user. On the other hand, the server should also raise its sending rate when bandwidth appears to be plentiful, and maximize the resource utilization and perceived quality. With the bandwidth estimation technique motivated by TCP Westwood (TCPW) [32], VTP satisfies all the above requisites. Details of VTP will be discussed in section 3.

In this work, we have implemented a fundamentally adaptive, end-to-end multimedia streaming system that allows a mobile user to receive uninterrupted service of best possible quality multimedia, while roaming among multiple heterogeneous wireless networks. Although the general concepts of providing adaptive services are not new, we aim to provide insights on end-to-end dynamics of such system from an implementation perspective instead of a simulated one. Actual system measurements collected from our testbed show that the combination of USHA and VTP can indeed provide substantial improvements to streaming performance, in terms of perceived video quality (smooth video frame rate), and robustness against sudden changes in link capacities.

The rest of this paper is organized as follows. Section 2 presents some background and discusses related work in the area of seamless handoff and video streaming. Section 3 describes the novel system unification of our seamless handoff architecture (USHA) and VTP. Section 4 presents actual end-to-end measurement results of the system from our Linux testbed. Section 5 concludes the paper.

2. Background and Related Work

2.1 Seamless Handoff

Handoff occurs when the user switches between different network access points. Handoff techniques have been well studied and deployed in the domain of cellular system and are gaining a great deal of momentum in the wireless computer networks, as IP-based wireless networking increases in popularity.

Differing in the number of network interfaces involved during the process, handoff can be characterized into either vertical or horizontal [30], as depicted in Figure 10.1. A vertical handoff involves two different network interfaces, which usually represent different technologies. For example, when a mobile device moves out of an 802.11b network and into a 1xRTT network, the handoff event would be considered as vertical. A horizontal handoff occurs between two network access points that use the same technology and interface. For example, when a mobile device moves between 802.11b network domains, the handoff event would be considered as horizontal since the connection is disrupted solely by the change of 802.11b domain but not of the wireless technology.

A seamless handoff is defined as a handoff scheme that maintains the connectivity of all applications on the mobile device when the handoff occurs. Seamless handoffs aim to provide continuous end-to-end data service in the

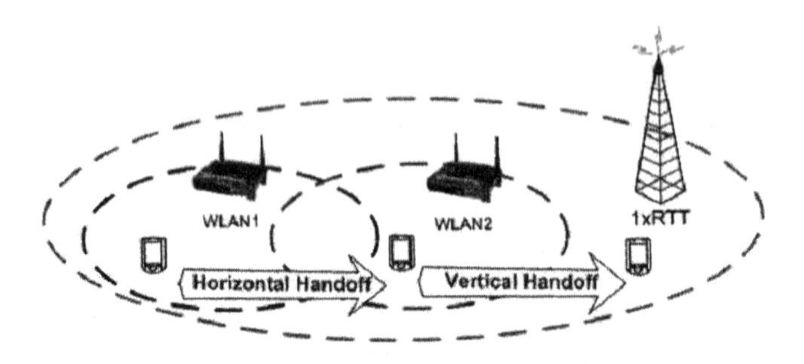

Figure 10.1. Horizontal and Vertical Handoff

face of any link outages or handoff events. Low latencies and few packet losses are the two critical design goals. Low latencies require that path switches be completed almost instantaneously; service interruptions should be minimized. In case of an actual connection failure, the architecture should attempt to reconnect as soon as the service becomes available; packet losses due to the switch should also be minimized.

Various seamless handoff techniques [9] [17] [19] [22] have been proposed. These proposals can be classified into two categories: network layer approaches and upper layer approaches. Network layer approaches are typically based on IPv6 [8] or Mobile IPv4 [21] standards, requiring the deployment of several agents on the Internet for relaying and/or redirecting the data to the moving host (MH). Most upper layer approaches implement a session layer above the transport layer to make connection changes at underlying layers transparent to the application layer [12] [15] [23] [27] [28]. Other upper layer approaches suggest new transport layer protocols such as SCTP [29] and TCP-MH [24] to provide the necessary handoff support.

Previous seamless handoff solutions, whether mobile IP based or mobile IP-less, are often elaborate to implement and to operate. For the network layer solutions, deployment translates to upgrading every existing router without mobile IP capabilities. The cost imposed by these solutions is an existing barrier to wide deployment. For the upper layer solutions, a new session layer or transport protocol calls for an update to all existing applications and servers not supporting it, the potential cost is also discouraging. Consequently, even though many handoff solutions have managed to minimize both latency and packet loss, they are often not deployed in reality by the majority of service providers. With the proliferation of mobile applications and mobile users, a "simple" and "practical" seamless handoff solution with minimal changes to the current Internet infrastructure remains necessary.

USHA, an upper layer solution providing simple and practical handoff solution, is deployed in our experiments to handle seamless vertical handoffs. Details of USHA will be presented in section 3.

2.2 Video Streaming

Multimedia streaming, in particular video, has been growing as an important application on the Internet. However, the Internet is inherently not appropriate for such applications. Unlike conventional data transfers such as FTP, for which the Internet was designed decades ago, streaming usually has more strict QoS constraints on delay, bandwidth, etc. The best-effort Internet architecture lacks built-in schemes to guarantee these constraints. Thus enormous efforts have been put into research on streaming over IP networks.

On the video compression side, popular standard algorithms such as MPEG-4 [26] and H.263 [18] produce encoded streams in a wide range of rates. On the networking side, the key issue is to estimate an eligible rate at which the server should send in order to maximize the utilization of network bandwidth while effectively sharing it with other flows. There are two classes of techniques to estimate eligible rates: with network feedback and end-to-end. On the Internet, due to various scalability and deployment issues, end-to-end techniques seem more practical.

Several solutions based on TCP congestion control have been proposed for the transport of video over the Internet. For instance, SCP [5], a TCP Vegas [4] -like rate adjustment method, suffers from the same problems as TCP Vegas, thus remains inherently unfriendly to other TCP flows in many scenarios. RAP [31], a protocol employing AIMD to adapt the sending rate as TCP, does not take retransmission timeout into account, and therefore may result in poor performance when the impact of timeout is significant. TFRC [10] is a popular equation based solution built upon the model of TCP Reno, and aims to provide good smoothness and TCP friendliness. However, efficiency of TFRC is susceptible to random losses at wireless links, a legacy problem from TCP Reno.

Additionally, TCP based streaming approaches also suffer delayed reactions to network dynamics in mobile scenarios (e.g. the maximum increase in TFRC's sending rate is estimated to be 0.14 packet/RTT and 0.22 packets/RTT with history discounting [10]). Consider a scenario where a video client handoffs from a low capacity link to a high one. A TCP based approach would use the "congestion avoidance" technique to linearly (and slowly) probe the available bandwidth on the new link. Such a slow reaction to network dynamics is unsatisfactory and can easily impair the overall experience of the client. As a result, a fast adaptive streaming technique is clearly a requisite for mobile needs.

Some of the commercial products claim to support adaptive video streaming, e.g. Helix Universal Server [16] and Microsoft Media Server [25]. However, lack of product disclosure and related analysis hinders independent efforts to verify the claims or to evaluate the streaming performance.

On the research side, some ongoing projects utilize packet pair/train measurements to estimate the end-to-end capacity and/or available bandwidth (or residual capacity), and adapt the sending rate accordingly. For example, inspired by TCP Westwood and its Eligible Rate Estimate (ERE) concept, SMCC [1] and VTP [2] are capable of adapting the sending rate to existing path conditions and resulting in both efficiency, i.e. high utilization of the bottleneck link, and friendliness to legacy flows. This enables faster responses in mobile handoff scenarios as well as achieving TCP friendliness. As a result of VTP's capabilities, it is used in this paper to evaluate the benefits of video adaptation in handoff scenarios. The VTP overview will be presented in section 3, and the experiments will be presented in section 4.

3. Proposed Approach

3.1 Universal Seamless Handoff Architecture

Universal Seamless Handoff Architecture (USHA), is a simple handoff technique proposed in [6] to deal with both horizontal and vertical handoff scenarios with minimum changes to current Internet infrastructure (i.e., USHA only requires deployment of handoff servers on the Internet.) USHA is a mobile IP-less solution; however, instead of introducing a new session layer or a new transport protocol, it achieves seamless handoff by following the middleware design philosophy [11], integrating the middleware with existing Internet services and applications. The simplicity of USHA makes it an attractive choice for a seamless handoff test bed.

USHA is based on the fundamental assumption that handoff, either vertical or horizontal, only occurs on overlaid networks with multiple Internet access methods (e.g. soft handoff), which translates to zero waiting time in bringing up the target network interface when the handoff event occurs. If coverage from different access methods fails to overlap (e.g. hard handoff), it is possible for USHA to lose connectivity to the upper layer applications.

In Figure 10.2, a handoff server (HS) and several mobile hosts (MHs) are shown. USHA is implemented using IP tunneling techniques (IP encapsulation), with the handoff server functioning as one end of the tunnel and the mobile host as the other. An IP tunnel is maintained between every MH and the HS such that all application layer communications are "bound" to the tunnel interface instead of any actual physical interfaces. All data packets communicated

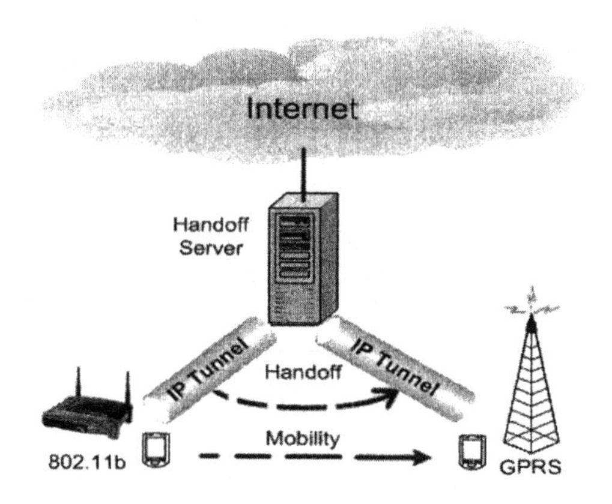

Figure 10.2. Universal Seamless Handoff Architecture

through this IP tunnel are encapsulated and transmitted using the connectionless UDP protocol.

The IP tunnel above utilizes two pairs of virtual/fixed IP addresses, one on HS and one on MH. The fixed IP addresses are necessary for an MH to establish a physical connection to the HS. When the handoff event occurs and the physical connection from MH to HS changes, the MH is responsible for automatically switching the underlying physical connection of the virtual tunnel to the new interface, as well as notifying the HS of its change in physical connection. Upon handoff notification, the HS immediately updates its IP tunnel settings so that any subsequent data packets will be delivered to MH's new physical link.

Since all data packets are encapsulated and transmitted using UDP, there is no need to reset the tunnel after the handoff. Therefore, end-to-end application sessions (e.g. TCP) that are bound to the IP tunnel are kept intact. This provides handoff transparency to upper layer applications.

A simple USHA testbed is implemented. Experiments and evaluation of adaptive video streaming in vertical handoff scenarios on this testbed will be discussed in section 4.

3.2 VTP

Bandwidth Estimation. VTP is a video streaming protocol aiming to adapt its rate and quality according to network conditions. The core of VTP is its bandwidth estimation technique. It estimates the Eligible Rate Estimate (ERE) by applying an Exponentially Weighted Moving Average (EWMA) to the achieved rate, which is in turn calculated as the number of bytes delivered to the client during a certain time interval, divided by the length of the interval.

Assume we use packet trains of length k to measure the achieved rate. Denote d_i as the number of bytes in packet i, t_i as the time when packet i arrives at the client. The sample of achieved rate measured when packet j is received, denoted as b_j, is

$$b_j = \frac{\sum_{l=0}^{k-1} d_{j-l}}{t_j - t_{j-(k-1)}} \tag{10.1}$$

The EWMA is needed to smooth achieved rate samples and eliminate random noise. Denote B_i as the available bandwidth estimate after getting sample b_j, then

$$B_j = \alpha \cdot B_{j-1} - (1 - \alpha)(\frac{b_j + b_{j-1}}{2}) \tag{10.2}$$

The reason of using both b_j and b_{j-1} is to further reduce the impact of randomness in the achieved rate samples.

Rate Adaptation. Current VTP implementation works with pre-stored streams but can be extended to live video. Multiple streams of the same content are encoded *discretely* at different rates. Compression algorithms such as MPEG-4 can adjust parameters, such as the Quantization Parameter (QP), to achieve different encoding rates. For example, a movie trailer may be encoded at 56Kbps, 150Kbps and 500Kbps, targeting users with different access capacities. VTP chooses from multiple encoding levels of the same content the best rate according to ERE. Figure 10.3 illustrates with a finite state machine how rate adaptation is performed in VTP. Three video encoding levels, namely Q0, Q1 and Q2 with ascending rates, are shown. IR0 through IR2 are the "increasing rate" states while DR is the "decreasing" rate state.

VTP starts from state Q0. Upon receiving an ACK from the client, VTP server compares its current sending rate with the recently updated bandwidth estimate B. If the sending rate is less than or equal to B, VTP regards it as an indication of good network condition and makes a transition to IR0, where VTP linearly increases its sending rate to probe the available bandwidth. The amount of rate increase is limited to 1 packet/RTT, same as in TCP. On exiting IR0, VTP may move to state Q1 when the rate is high enough to support the level 1 stream, i.e. quality upgrade; or return to Q0 otherwise. Thus Q0 only implies the server is sending the level 0 stream; it says nothing about the actual sending rate. This process repeats itself, with possible quality upgrades, until the bandwidth estimate drops below current sending rate.

Rate decrease happens immediately when the measured bandwidth estimate drops below the sending rate. A transition from the current encoding level, say

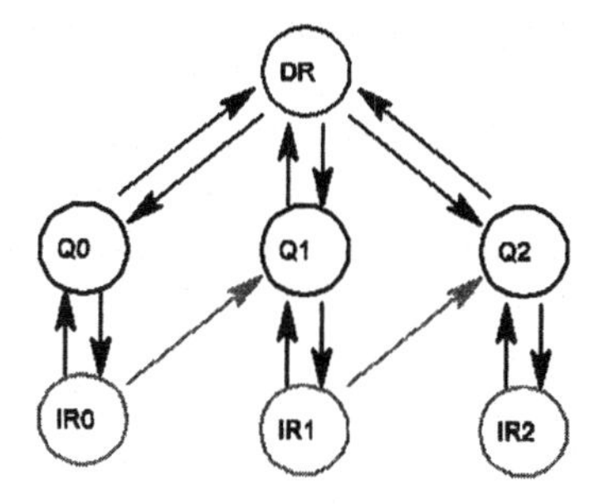

Figure 10.3. Rate adaptation in VTP

Q2, to DR is made. In DR, sending rate is decreased to the bandwidth estimate. If this rate is no longer able to support the current encoding level (level 2 in this example), one or more level decreases, i.e. quality downgrade, will occur until the level that the new sending rate can support. If the sending rate is below Q0, the lowest level, the streaming service will either be stopped or send at this lowest level, depending on administration policies.

Transmission Scheduling for VBR Video. Constant Bit Rate (CBR) video continuously adjusts QPs to maintain the target bit rate of the stream. This simplifies transmission scheduling, but results in varying video quality from frame to frame, which is unpleasant to the viewer. On the other hand, Variable Bit Rate (VBR) video produces streams with varying bit rates; and with more consistent quality.

Due to space limit we will not cover VTP transmission scheduling in detail in this paper. Briefly speaking, VTP divides a video clip into a number of segments. For each segment, VTP computes a target rate, at which neither buffer overrun or underrun should occur. Since video streams are pre-stored, instantaneous sending rates are available beforehand, and so are the target rates of the segments. VTP then applies these target rates to the finite state machine in Figure 10.3 for rate adaptation.

In the next section we will evaluate the performance of adaptive video streaming in seamless handoff scenarios of our integrated USHA + VTP testbed.

4. Experiments

In this section, we present measurement results of adaptive video streaming in vertical handoff scenarios using a 2-minute movie trailer encoded in MPEG-4 at three discrete levels. We denote them as levels 0, 1, and 2, corresponding to the encoding rates (VBR) of below 100, 100 $\sim$ 250, and above 250 Kbps, respectively. The VTP server is implemented on a stationary Linux desktop; the client is on a mobile Linux laptop. The USHA system is also set up in Linux, with custom configured NAT and IP tunneling. Both the VTP server and client are connected to the handoff server, the former via 100 Mbps Ethernet; the later via 802.11b and 1xRTT provided by Verizon Wireless. The 802.11b is set at the 11 Mbps mode; the bandwidth of 1xRTT varies with cross traffic, the typical value is around tens of Kbps.

We have tested two handoff scenarios, from 1xRTT to 802.11b (low capacity to high) and vice versa. In all experiments, one-time handoff occurs at 60 sec after the start of the experiment. In each scenario, we have tested both non-adaptive and adaptive video streams. In the non-adaptive case, video of fixed quality is sent throughout the experiment regardless of ERE, while in the adaptive case the video quality adapts accordingly.

4.1 Handoff from 1xRTT to 802.11b

In the first set of experiments, we evaluate the performance of video streaming when the mobile host performs handoff from the lower-capacity interface of 1xRTT to the higher-capacity interface of 802.11b.

Non-adaptive Video Streaming. First, we run "non-adaptive" experiments one for each encoding level. Since the coding rates of levels 1 and 2 are both above the capacity of 1xRTT, the corresponding experiments "died" shortly after started simply because of the inability of 1xRTT to handle such high rates. Results are not reported. Only video of level 0 made it through as the results show below. More specifically, Figure 10.4 shows the frame rate received by the mobile client, and Figure 10.5 shows the sending rate at the VTP server. In Figure 10.5, "Reference Rate" means the source rate of the video stream (note that the source rate is variable, even within a given encoding scheme), whereas the "Sending Rate" means the instantaneous transmission rate of the data, which depends on the link capacity and thus may exceed the source rate.

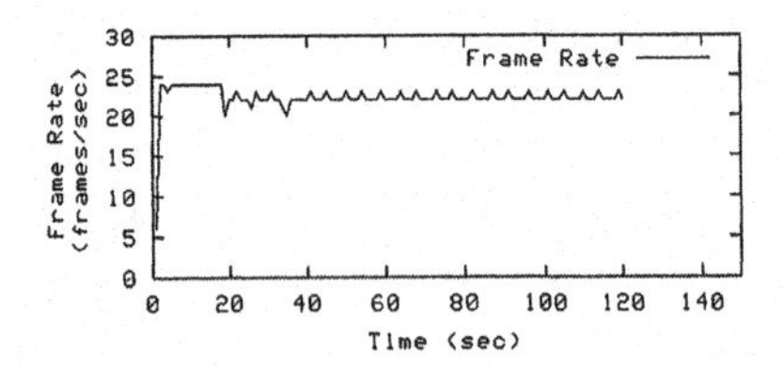

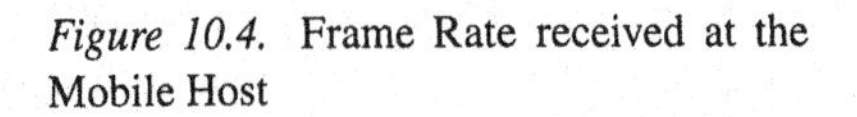
Figure 10.4. Frame Rate received at the Mobile Host

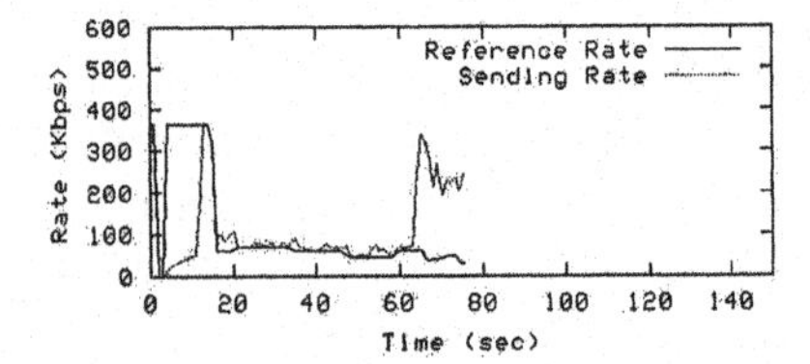

Figure 10.5. Sending Rate at the Video Server

In Figure 10.4, the video frame rate is stable and consistently between a visually pleasing range of 20 and 25 frames/sec (fps) shortly after it is started. Even in the presence of a handoff from LOW to HIGH at 60 sec, the frame rate remains unaffected. This proves our USHA to be transparent to applications. The video quality is overall very good in terms of smoothness. However, Figure 10.5 reveals more insightful information. In this non-adaptive experiment, the reference rate and video quality remain low after the handoff at 60 sec, where they could increase to take the advantage of the increased "sending rate" and bandwidth. This justifies the exploration of adaptation in video streaming applications. Note that after the handoff, the actual sending rate is much higher than the reference rate, so the server finishes sending quickly (before 80 sec).

Adaptive Video Streaming. The setup of adaptive streaming experiment is similar to the non-adaptive one described above except that now the video quality level adapts to the network conditions. In Figure 10.6, we show the frame rate received by the mobile client. Still it is stable and consistently in a range that gives good perceived quality. No dips in frame rate are found when the handoff event occurs.

Figure 10.7 shows the quality level of the video sent by the VTP server (averaged over 1-sec intervals). Level 2 is highest and 0 is lowest. Prior to the handoff at 60 sec, most frames are sent at the lowest quality level (i.e. 0); after handoff the average quality jumps to about 1.5. This is consistent with our experiment setup where the available bandwidth increases drastically when moving from 1xRTT to 802.11b.

Figure 10.8 shows the reference and sending rates on the VTP server. Prior to the handoff at 60 sec, Figure 8 looks very similar to Figure 10.5. The difference emerges after the handoff. The reference rate jumps up and strives to match the sending rate (300 Kbps), indicating that high quality video is now being transmitted across the 802.11b channel. In other words, VTP successfully detects (within fractions of a second) the change in available bandwidth and adapts its video encoding level to maximize the perceived quality of the mobile user.

4.2 Handoff from 802.11b to 1xRTT

In the second set of experiments, we evaluate the performance of video streaming when the mobile host performs handoff from the high-capacity interface of 802.11b to the low-capacity interface of 1xRTT. To make results comparable to the previous experiments, the one-time handoff is also generated 60 sec after the experiment is started.

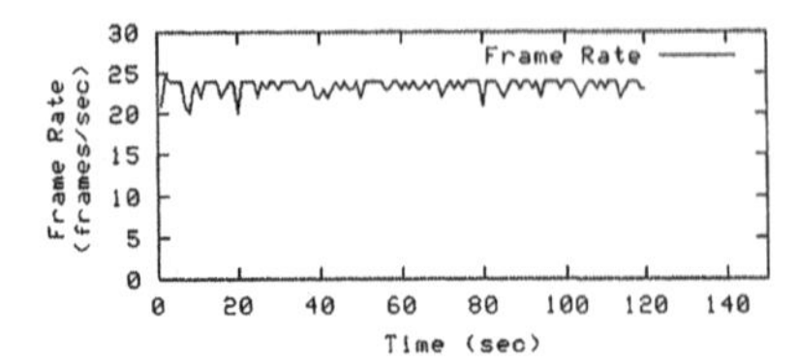

Figure 10.6. Frame Rate received at the Mobile Host

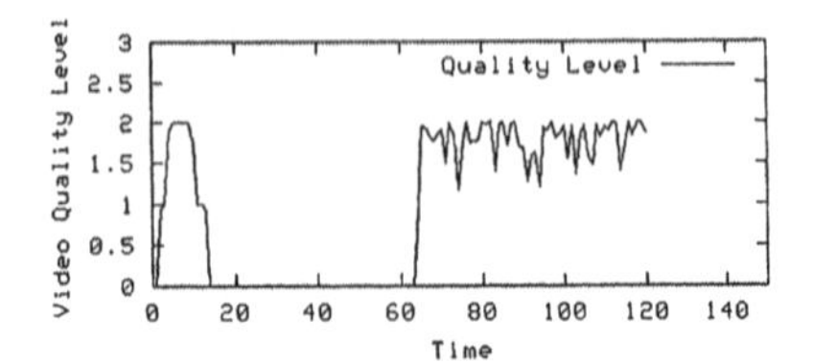

Figure 10.7. Video Quality sent at the Video Server

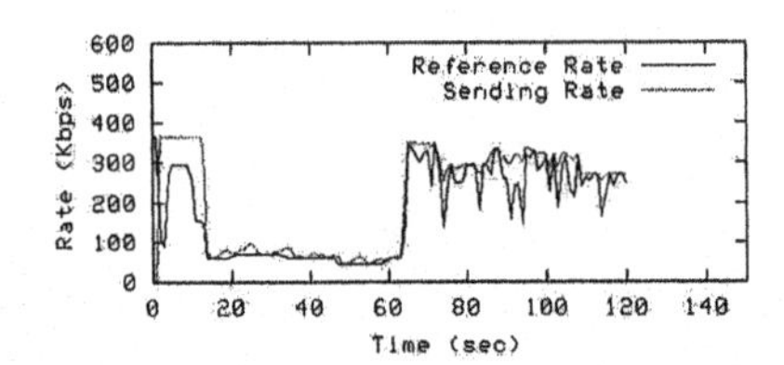

Figure 10.8. Sending Rate at the Video Server

Non-adaptive Video Streaming. Similar to the experiments that we have done in the case where handoff occurs from 1xRTT to 802.11b, we have also tested non-adaptive streaming at all three quality levels, respectively. Unlike the previous experiments, this time handoff occurs from the high-capacity interface to the low-capacity one, thus all three levels are feasible initially and can be tested. As expected, after the handoff, experiments with levels 1 and 2 virtually "died".

We show the experiment results with level 2, i.e. the highest quality in Figure 10.9 (video frame rate received by the mobile client) and Figure 10.10 (sending rate on the VTP server). Before the handoff, the frame rate received by the client is high and stable, and the reference and sending rates at the server are both high and close to each other, an obvious sign of high quality video. These metrics drop sharply at 60 sec when the handoff occurs, the reason being that 1xRTT is not able to handle the video of highest quality as we have explained. The frame rate drops to an unacceptable level of 10 fps; the sending rate becomes less than half of the reference rate. In the experiment we have found that the video virtually "froze" after the handoff. This experiment confirms the claim that adaptive multilevel video codes are a must in heterogeneous roaming.

Adaptive Video Streaming. Moving on to the adaptive video experiments, Figure 10.11 shows the video frame rate received at the mobile client - high and stable as we have seen in Figure 10.6. Note that there exists a small dip in the

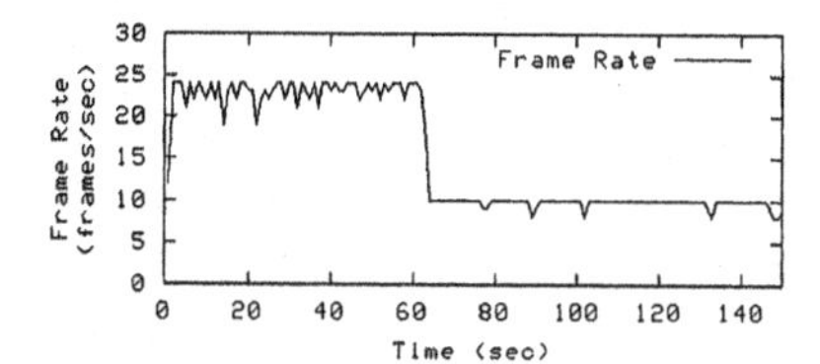

Figure 10.9. Frame Rate received at the Mobile Host

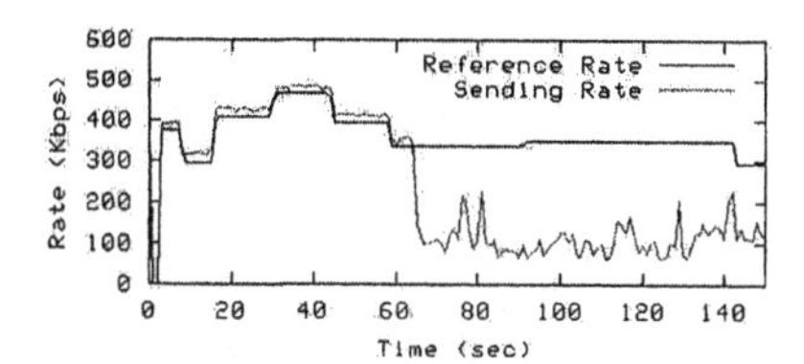

Figure 10.10. Sending Rate at the Video Server

frame rate shortly after the handoff event at 60 sec, but the recovery is within a couple of seconds. This again proves the effectiveness of seamless handoff and rate adaptation of our proposed solution.

Figure 10.12 and Figure 10.13 show the video quality level and the reference and sending rates at the VTP server. Prior to the handoff at 60 sec when the system is running over the 802.11b connection, video quality is high (i.e. 2), so is the reference rate, matching the sending rate. Exactly at 60 sec the system is able to detect the handoff event and to adapt the video quality to the reduced bandwidth. Throughout the experiment the sending rate is always ahead of the reference rate so that there is no backlog build up at the sender.

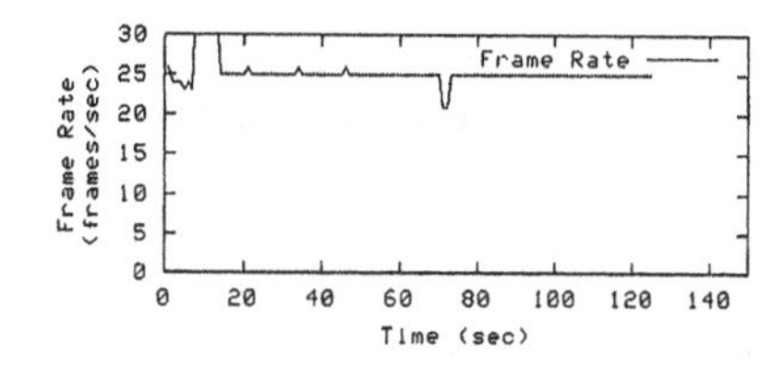

Figure 10.11. Frame Rate received at the Mobile Host

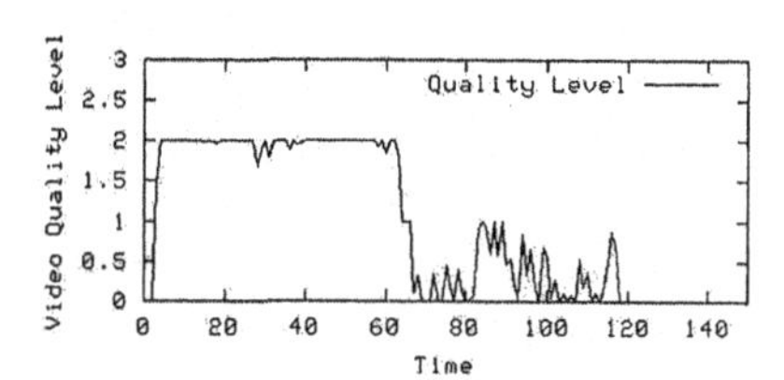

Figure 10.12. Video Quality sent at the Video Server

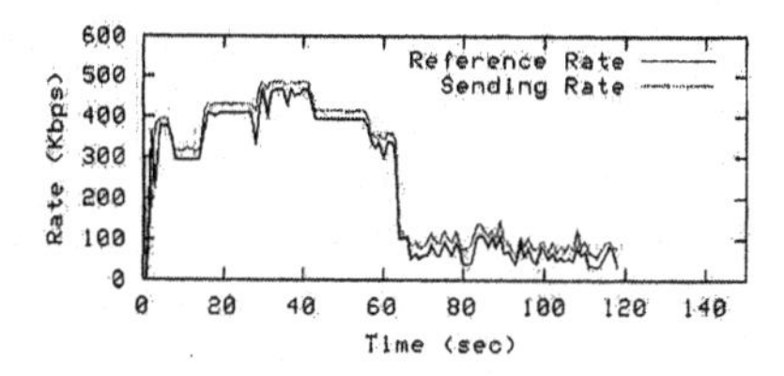

Figure 10.13. Sending Rate at the Video Server

4.3 Discussions

In a handoff-enabled environment, drastic changes in the link capacity are often associated with vertical handoff events. For instance, handoff from 1xRTT to 802.11b can easily witness a 100-fold increase in the link capacity (from 100 Kbps to 11 Mbps). Some traditional approaches (e.g. TFRC) incorporate the well-known *slowly-responsive congestion control* (SlowCC) [3] and thus can smoothly adjust the sending rate. However, SlowCC cannot take aggressive advantage of the rapid change of resources in emerging vertical handoff scenarios [13]. In order to utilize the bandwidth resources well, and maximize the user-perceived quality, a well-designed adaptive streaming scheme must take into account the effect of drastic capacity changes in both up and down directions.

From the experiment results presented in above, it is evident that VTP is one such scheme. Using the eligible rate estimate, VTP can properly and rapidly adapt its sending rate and video quality to available bandwidth, and hence is successful in handling vertical handoffs. This is not small feat. In fact, in most AIMD-based streaming protocols inspired to TCP, the adaptation process adjusts slowly to capacity changes. For example, when handoff occurs from LOW to HIGH (i.e. 1xRTT to 802.11b), no congestion loss is detected. A TCP based scheme will remain in congestion avoidance and linearly increase its congestion window (and rate) to probe the available bandwidth.

In the opposite direction, where handoff occurs from high (e.g. 802.11b) to low capacity (e.g. 1xRTT), there is immediate packet loss at the moment of the handoff, so AIMD protocols will react promptly to such loss. In fact, they tend to overreact causing oscillatory behavior and slower convergence to the new (lower) encoding rate.

In general, application performance would benefit if the server could predict the imminent handoff (e.g. MAC layer feedback from fading signals of one connection and strengthening signals of the other) and thus slow down its sending rate just before the handoff. We plan to address this issue in our future work.

5. Conclusion

In this work, we have studied the need and evaluated the performance of adaptive video streaming in vertical handoff scenarios. We have proposed an integrated solution of seamless handoff and adaptive video streaming, and implemented it on a Linux testbed, consisting of a USHA server and a VTP streaming system. Experiments on both non-adaptive and adaptive video applications, with handoffs from 1xRTT to 802.11b and vice versa, have been carried out to evaluate the performance of our proposed solution. From the measurements results we have seen that the USHA/VTP solution can effectively hide handoff events from the application and provide uninterrupted transport and application sessions during handoffs. Moreover, the adaptive streaming system is able to detect available bandwidth changes and adjust the video quality and sending rate accordingly. In summary, such a combination of adaptive video streaming and seamless vertical handoff will become very desirable in the emerging ubiquitous mobile computing environment.

References

[1] Aboobaker, N. et al. " Streaming Media Congestion Control using Bandwidth Estimation," in Proc. of IFIP/IEEE International Conference on Management of Multimedia Networks and Services, 2002.

[2] Balk, A. et al. "Adaptive Video Streaming: Pre-encoded MPEG-4 with Bandwidth Scaling,"" submitted for publication. www.cs.ucla.edu/NRL/hpi/tcpw/tcpw_papers/mpeg-cnj2003.pdf

[3] Bansal, D. et al. "Dynamic Behavior of Slowly-Responsive Congestion Control Algorithms," In Proc. of ACM SIGCOMM 2001.

[4] Brakmo, L., O'Malley, S., and Peterson, L. "TCP Vegas: New techniques for congestion detection and avoidance," in Proc. of the SIGCOMM '94 Symposium (Aug. 1994) pages 24-35.

[5] Cen, S., Pu, C., and Walpole, J. "Flow and Congestion Control for Internet Streaming Applications," Proc. of Multimedia Computing and Networking, 1998.

[6] Chen, L.J. et al. "Universal Seamless Handoff Architecture in Wireless Overlay Networks," Technical Report TR040012, UCLA CSD, 2004.

[7] Davies, N. et al. "Supporting adaptive video applications in mobile environments," IEEE Communications Magazine, 36 (6):138–143, June 1998.

[8] Deering, S., and Hinden, R. "Internet Protocol, Version 6 (IPv6) Specification," RFC 2460, Dec. 1998.

[9] Dommety, G. et al. "Fast Handovers for Mobile IPv6," draft-ietf-mobileip-fast-mipv6-04.txt, IETF Internet draft, Mar. 2002.

[10] Floyd, S. et al. "Equation-Based Congestion Control for Unicast Applications," In Proc. of ACM SIGCOMM 2000.

[11] Fox, A. et al. "Cluster-based Scalable Network Services," SOSP'97, 1997.

[12] Ghini, V. et al."Smart Download on the Go: A Wireless Internet Application for Music Distribution over Heterogeneous Networks," in Proc. of ICC, 2004.

[13] Gurtov, A., and Korhonen, J. "Measurement and Analysis of TCP-Friendly Rate Control for Vertical Handovers," submitted for publication, http://www.cs.helsinki.fi/u/gurtov/papers/vho.html

[14] Handley, M. et al. "TCP Friendly Rate Control (TFRC): Protocol Specification," RFC 3448, January 2003

[15] Handley, M. et al. "SIP: Session Initiation Protocol," RFC 2543, March 1999.

[16] Helix Universal Server, http://www.realnetworks.com

[17] Hsieh, R., Zhou, Z. G., and Seneviratne, A. "S-MIP: a seamless handoff architecture for mobile IP," in Proc. of IEEE INFOCOM 2003.

[18] ITU-T Recommendation H.263, Feb. 1998

[19] Johnson, D. B., Perkins, C., and Arkko, J. "Mobility Support in IPv6," draft-ietf-mobileip-ipv6-17.txt, IETF Internet draft, May 2002.

[20] Lu, S., Lee, K.-W., and Bharghavan, V. "Adaptive service in mobile computing environments," in Proc. of IWQOS 1997, New York.

[21] Perkins, C. Ed. "IP Mobility Support for IPv4," RFC 3344, Aug. 2002.

[22] K. El Malki et al, "Low Latency Handoffs in Mobile IPv4," draft-ietf-monileip-lowlatency-handoffs-v4-03.txt, IETF Internet draft, Nov. 2001.

[23] Maltz, D., and Bhagwat, P. "MSOCKS: An architecture for transport layer mobility," In Proc. of IEEE Infocom, p.p. 1037-1045, March 1998.

[24] Matsumoto, A. et al. "TCP Multi-Home Options," draft-arifumi-tcp-mh-00.txt, IETF Internet draft, Oct. 2003.

[25] Microsoft Media Server, http://www.microsoft.com/windows/windowsmedia

[26] MPEG-4 Industry Forum, http://www.m4if.org

[27] Schlaeger, M. et al. "Advocating a Remote Socket Architecture for Internet Access using Wireless LANs," Mobile Networks & Applications, vol. 6, no. 1, pp. 23-42, January 2001.

[28] Snnoeren, C. "A Session-Based Approach to Internet Mobility," PhD Thesis, Massachusetts Institute of Technology, December 2002.

[29] Stewart, R. et al. "Stream Control Transmission Protocol," RFC 2960, Oct. 2000.

[30] Stemm, M. and Katz, R. H. "Vertical Handoffs in Wireless Overlay Networks," ACM MONET, 1998.

[31] Rejaie, R., Handley, M., and Estrin, D. "RAP: End-to-end Rate Based Control for Real Time Streams in the Internet," In Proc. of IEEE INFOCOM'99, 1999

[32] Wang, R., Valla, M., Sanadidi, M. Y., and Gerla, M. "Adaptive Bandwidth Share Estimation in TCP Westwood," In Proc. of IEEE Globecom 2002.

Chapter 11

ON SENSOR NETWORKING AND SIGNAL PROCESSING FOR SMART AND SAFE BUILDINGS*

Pramod K. Varshney and Chilukuri K. Mohan
Department of Electrical Engineering and Computer Science
Syracuse University, Syracuse, NY 13244
{varshney,ckmohan}@syr.edu

Abstract The monitoring and control of indoor environments has become an increasingly important concern, with multiple objectives including maximizing the comfort and minimizing health risks of occupants, in everyday operating conditions as well as emergency situations. The deployment of many networked sensors, equipped with reasoning and communicating abilities, can significantly contribute towards these objectives. This chapter discusses a framework for sensor networking in smart and safe buildings, and considers various issues relevant in the development of such networks and effective use of the information obtained by sensors in these networks. A hierarchical decision model is envisaged spanning multiple reasoning levels. Examples of control and response planning applications are discussed.

Keywords: Sensor networks, built environments, monitoring, control, hierarchical reasoning, information fusion.

Introduction

There is a rapidly growing interest in both the theory and practical applications of pervasive sensor networks and computing. The vision is to instrument our living environment at various scales and for a variety of scenarios with a

*We gratefully acknowledge the contributions of our colleagues Biao Chen, Can Isik, Ez Khalifa, Kishan Mehrotra, Lisa Osadciw and Jensen Zhang to the work described in this paper.

large number of sensors and actuators to monitor and control it in a manner so that it is customized for an individual or a group according to their preference. This vision can become a reality due to the tremendous advances made in the areas of miniaturized, low-power and inexpensive sensors and actuators, wireless networking and computing technologies. A broad spectrum of applications is being considered that include battlefield surveillance for the military; environmental monitoring such as for habitat, forests and water bodies; remote monitoring of patients; industrial applications such as monitoring of factories and appliances; monitoring of critical infrastructure such as bridges, dams and power grids; and environmental control within buildings. This evolving area has given rise to many challenging issues that need to be resolved before the vision can be realized. Research challenges include fundamental problems of networking, collaborative signal processing, information fusion, and control as well as application-specific issues of interoperability and performance evaluation. The goal is to look at each application at the system level and obtain solutions for different requirements and at different scale.

Each envisaged application is expected to touch people's lives and enhance the quality of life. In this paper, we focus on the application of sensor networking inside buildings to monitor and control the indoor environmental quality (IEQ) for different objectives and requirements and at different scales. This is a major societal issue that touches everyone's life and improvements in the IEQ by the use of sensor networking will have a major impact on the quality of life. There are two important and very timely applications of buildings instrumented with sensors and associated infrastructure. One involves emergency situations in combating terrorist attacks involving chemical/biological agents in a building, and the other is of a more routine nature in making indoor environments healthier for occupants. Both of these applications have much in common: they require sensor based monitoring, information transportation over reliable networks, intelligent signal processing followed by formulation of control and response plans. They also have some differences with respect to the objective functions to be optimized, control methodologies, response plans, time deadlines and criticality.

The indoor environment is critical to human health and quality of life. In the US, it has been estimated that people spend 90% of their time indoors. Also, 41% of homes, 33% of commercial buildings, and 58% of schools have inadequate indoor environmental quality (IEQ). Particulate matter is considered to be the leading cause of IEQ problems that lead to school absences and affect productivity in the workplace. It is estimated that 17.7 million people in the US suffer from Asthma, out of which 4.8 million are children. Productivity loss in the US has been estimated to be as high as $250B due to poor IEQ. Therefore, it is imperative that IEQ be monitored and controlled in indoor environments by introducing chemical sensors to augment the sensors commonly found in the buildings such as the thermal sensors. Most of the work thus far [3, 1] in the areas of intelligent and high performance buildings has focused on improving comfort level and energy conservation. Recent advances in sensor networks and associated information processing techniques when deployed in smart buildings can enable the control of IEQ in a comprehensive manner for improved human health and productivity at reduced energy costs.

In addition to the routine operations of buildings, another scenario has emerged that requires immediate action and an instrumented building is highly essential for this application. As the relatively recent tragic events of Anthrax attacks have clearly demonstrated, occupants are quite vulnerable to chemical and biological attacks, particularly if released inside buildings. The most dangerous forms of Chemical and Biological Agents (CBA) are gaseous chemicals and very fine chemical or biological aerosols that could be easily dispersed in the air and remain airborne for a very long time. Fine aerosols are also more difficult to remove by filters and eliminators, including human body filtering mechanisms. CBA could be released intentionally or accidentally inside a built environment (buildings, trains, cars, etc). This release of CBA poses a major challenge that necessitates quick responses and countermeasures. For example, in case of a terrorist attack it is extremely important to maximize the probability of detection of CBA at the possible expense of a few false alarms; decisions will have to be made very quickly and the response plan is expected to involve decontamination and evacuation. The challenges include the ability to detect a chemical and biological release, localize its source, and differentiate the nature of the released agent, determine or predict its spatial and temporal dispersion pattern, and decide and communicate quickly and efficiently during and after such attack. Availability of a sensor network for information acquisition and processing will help immensely in developing approaches for defense against such attacks and for evacuation of occupants from contaminated buildings, thus making the buildings safer for occupants.

The rest of the paper is organized as follows. In the next section, we provide a high-level description of the sensor networking problem for the intelligent building application. Associated research issues are also discussed. Section 2 discusses characteristics of sensors and their attributes relevant to the building environment application. Sensor network deployment and operational issues are discussed in Section 3. Considerations relevant to the analysis of data obtained by sensors are briefly discussed in Section 4; these include control and response planning issues. Two examples of the same, including some recent results, are discussed in Section 5. Concluding remarks are provided in Section 6.

1. A Framework for Sensible Building Environmental Systems

Built environmental systems (BES) include buildings, cars, airplanes, ships, spacecraft, and other structures that provide desired environmental conditions for human occupants. Smart sensor based systems may be envisioned for environmental control for each one of these applications; in this paper, we focus only on buildings.

A building environmental system consists of several sub-systems: building envelope, HVAC (Heating, Ventilation and Air-Conditioning), lighting, power and energy, information and communication, fire and safety, water supply, and drainage systems (see Figure 11.1). Each sub-system interacts with the environment outside the building through flows of heat, moisture, contaminants, light, sound, water, power and energy, data and information. Humans interact with building systems through generation of heat and contaminants, and management and control of system operations. In addition to building environmental systems, weather conditions and the environment surrounding the building also affect IEQ. For example, air pollution and noise can significantly affect a building's air and acoustical quality. In fact, a building can be viewed as a nested multi-scale system involving a personal micro-environment surrounding a person, individual room environment, and the whole building environment. By employing dense sensor networks, information at various scales can be acquired and customized control strategies can be applied at each scale for achieving optimal IEQ, high energy efficiency and adequate building security.

The overall system will rely on sophisticated sensor-based systems. A large number of sensitive, real-time sensors, preferably agent-specific, would be distributed in the building, particularly at the most likely release or early detection sites. The nature of the sensors, the manner in which they are networked, and other details of the intelligent processing system depend on the constraints, goals and requirements for each specific building, which may include the following:

Safety: The presence of lethal chemical/biological agents needs to be detected rapidly and immediate actions taken to ensure minimizing the hazard to humans.

Health: Actions need to be taken when concentration levels of certain chemicals in the air exceed prespecified thresholds.

Comfort: The personal environments of various humans in the buildings need to be matched with actual or predicted requirements of individual users.

Energy: The overall energy consumption levels associated with the building must be reduced.

Cost: Infrastructural, maintenance, and operational costs must be reduced.

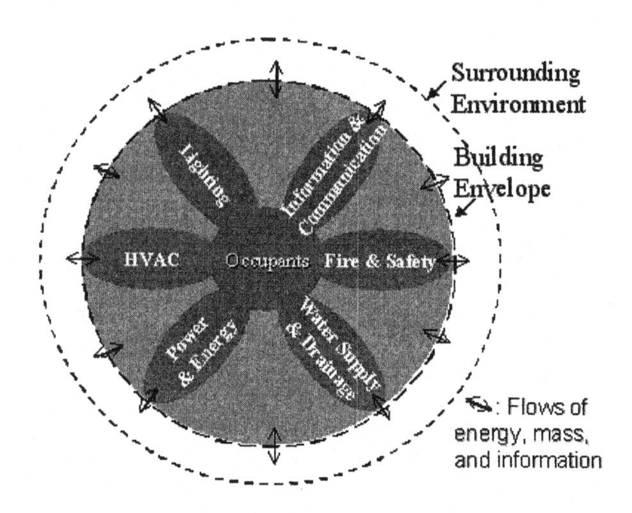

Figure 11.1. Interacting Entities in a Building Environment

Security: The quality of the built environment must not be susceptible to attack by malicious agents who gain access to a subset of computing, communication, or sensing components of the system.

Robustness: System performance should not be affected drastically by the failure of a small number of sensors or other components.

Modifiability: It should not be difficult to add or delete sensors or other components into the system, or to interact with new components about which prior information is not available when the system is first established.

The relative importance of these considerations will vary for specific systems. However, a broadly applicable perspective of the overall system can be visualized in terms of the conceptual process diagram shown in Figure 11.1.

In this framework, information regarding the indoor environment is to be collected by a number of strategically placed and often redundant sensors that measure a variety of environmental parameters and conditions. This information is transported over a reliable communication network to computing nodes, possibly connected by wireless links. This network should be secure, to provide assured information transportation services and to protect the information infrastructure from cyber attacks. We envisage a hierarchical distributed computing framework (described later) so that computing nodes at different levels have different capabilities and responsibilities. At the computing nodes, in-

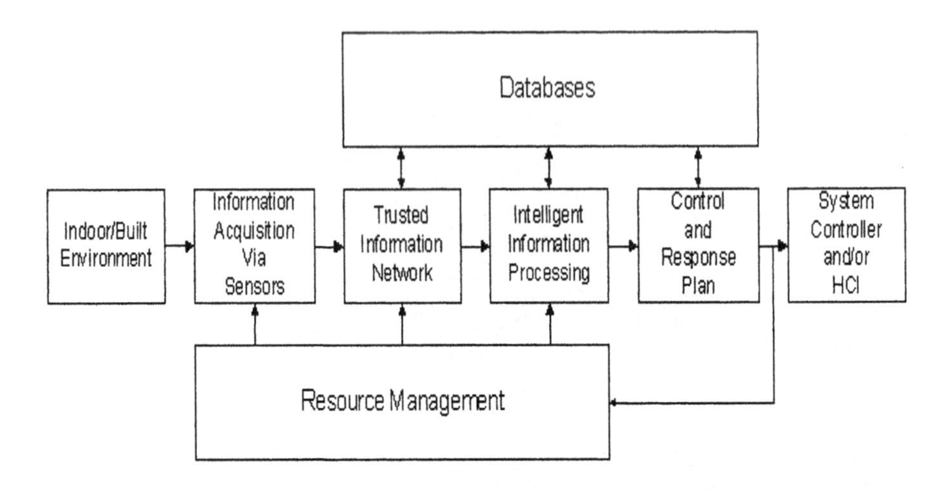

Figure 11.2. Conceptual diagram for the overall system for IEQ monitoring and control

formation is processed using methods and tools from the research areas of information fusion, knowledge discovery, and optimization.

Control and response plans are determined based on the information received and processed. These actions may include changing equipment settings and/or activating additional equipments, or developing decontamination and evacuation plans in case of an emergency. These plans are conveyed to a system controller or the human computer interface so that appropriate actions can be taken. Many of the system blocks in Figure 11.2 interact with databases while carrying out their functions. The proposed system also contains a resource control mechanism that influences operations from information acquisition to processing. For example, additional information may be collected, or a different information processing algorithm may be employed under direction from the resource controller. Later sections discuss some of the components of the above system in more detail.

2. Sensor Modeling

Sensors for the indoor environment may be characterized based on the following features:

Target: Most sensors will be specifically targeted towards sensing a single contaminant molecule or environmental feature (such as temperature or humidity). Some sensors can also be sensitive to a class of chemical compounds rather than a single molecule.

Response curves: Sensors vary in their outputs when presented with different concentrations or levels of the stimulus. A single "step function" response may be expected in some cases, i.e., the output of the sensor indicates whether or not the contaminant concentration exceeds a certain threshold. S-shaped sigmoid functions are also possible, with hysteresis loops. Another class of response curves is monotonic, e.g., the magnitude of sensor output is roughly linearly proportional to the sensed concentration.

Robustness: Some sensors exhibit almost the same response curves under a large variety of environmental conditions, whereas others are affected by conditions such as humidity and ambient temperature.

Mobility: The sizes of some sensors are small enough to permit considerable movement; such sensors may be allowed to detect concentration gradients and "migrate" towards regions of higher concentrations of sensed pollutants. Even for more conventional sensors, it is possible to provide a limited degree of freedom in movement (e.g., using tracks on the ceiling or along the walls) to permit more accurate detection of contaminant concentration levels and to pinpoint source locations.

Cost: Eventual deployment of sensors will be governed by economic considerations, weighing the costs of the sensors and associated infrastructure against the potential benefits to be obtained or the potential calamities to be avoided. Large medical, military and high security buildings will hence be the first candidates for deployment of intelligent sensor networks. Tradeoffs exist, e.g., infrastructure costs in increasing sensor mobility may achieve results equivalent to having a larger number of immobile sensors. Similarly, a sensor with lower cost and lower sensitivity may be preferred to one with higher cost and higher sensitivity in less critical building environments.

At any given moment, each sensor that is not disabled may be in one of two states: Dormant, or Active. A dormant sensor is currently inactive (i.e., not sensing or communicating), but may be rendered active as a result of a specific external event such as the elapse of prespecified time interval, or a received electrical signal. Active sensors may be sensing the environment (i.e., receiving external inputs and generating outputs), moving (if mobile), communicating, or a combination of these three activities. Communicating sensors may be receiving or sending messages from/to specific nodes or all accessible communicating sensors. Sensors may be structured into a hierarchy of clusters based on physical localization, and multicast operations may be defined to communicate with all nodes in the cluster (containing the communicating node) at any level.

3. Sensor Network Deployment and Operation

A reliable and efficient communication infrastructure is indispensable for the implementation of the indoor intelligent sensor network. Although communication between sensor-level intelligent subsystems and the central units may use wired or wireless media, wireless sensor networks may be preferable for the following reasons:

1. Cost-effective implementation of wireless networks would be easier in existing buildings without prior infrastructure.

2. Wireless networks facilitate rapid installation, reconfiguration, and adaptation, e.g., when structural changes are made to a building, or when a new threat is perceived, or when new sensors need to be placed.

One may consider existing indoor wireless data (non-voice) networks include wireless LAN as well as ad hoc networks such as Bluetooth. The latter are crippled by limited transmission range, and by the difficulty of global management needed for secure and effective functioning of a hierarchical system. Wireless LAN models such as those specified in IEEE 802.11a [8] and HIPERLAN/2 [4] support multi- rate communications and may provide a viable option for the IEQ application.

Network management and resource allocation is more complicated in a network for building applications than a commercial wireless network due to a variety of reasons [12]. In commercial wireless networks, communication is sporadic and user initiated [6]. In a wireless sensor network, data needs to be collected at synchronized time intervals and analyzed collectively. Also, the sensor suites are heterogeneous and irregularly organized due to performance and security requirements. For example, measurements of temperature, humidity, and detection of chemical and biological agents require very different sensors and measurements. Security concerns add to the complexity, requiring network adaptability in the event of a compromised sensor or communication links. Finally, energy efficiency of the communication links is also a factor for micro-sensors in the network [7]. These and several other issues such as quality of service (QoS) need further investigation for the specific application of IEQ monitoring and control.

Sensor networks such as those envisaged here also raise important maintenance and operational issues. In particular, since emergency situations rarely arise (by definition), the failure of a sensor or its associated components (e.g., for communication) may go undetected for a long time, inhibiting the ability of an organization to respond effectively when an emergency does arise. This calls for periodic testing as well as robustness in design of the network, so that the failure of a small number of sensors or associated components does not cripple the entire system. As the sensor network is being designed, simulations may be carried out to determine potential trouble spots if a few sensors fail. Redundancy in coverage may be built into the network, although this comes at the additional cost involved in deploying more sensors.

4. Data Analysis and Response Planning

A key design decision involves determining the amount of "intelligence" associated with each sensor. At one extreme, sensors may communicate their raw numerical outputs to centralized processors; this considerably increases bandwidth requirements and has the potential of clogging the centralized processors with too much data at critical instants of time when the processor needs to be involved in other activities. At the other extreme, data analysis as well as initiation of control actions may be completely decentralized and activated by the outputs obtained from individual sensors; this requires the formulation of protocols to address possible conflicts in the inferences made at different sensors, as well as the possibility of incorrect operation of some sensors. Intermediate solutions may be best, in which each sensor has a limited amount of data analysis capability, and communicates limited information (possibly only under situations requiring drastic control action) to higher level nodes. Decisions or control actions that may affect zones of the building sensed by multiple sensors (with overlapping regions of sensing) ought to be made by such sensors communicating and acting in concert, possibly through the intervention of a local "manager" node or processor. The primary activities of the decision-making nodes include:

Information Fusion: Drawing inferences from the outputs communicated by multiple sensors.

Predictive analysis: Making inferences regarding the expected future state of the building environment, based on present and past data, and relying on physical airflow models.

Control and response planning: Determining the actions needed to optimize human health, comfort, and related concerns.

The following subsections discuss the decision-making model and control issues.

4.1 Hierarchical Decision-Making Model

In the prototype framework discussed in Section 1, the indoor environment is monitored by a large number of "intelligent" sensor suites placed in the building and organized as a hierarchical network. This calls for the design of a distributed multi-agent system in which each sensor is capable of reasoning and communication. The architecture of such a system is proposed to be hierarchical and heterogeneous, with agents at different levels of the hierarchy being capable of different degrees of intelligence and imbued with different degrees of control over other sensors. Since this application requires fast response time, each sensor will be equipped with sufficient intelligence to permit it to sense the environment, perform elementary analysis of such inputs, and communicate with other intelligent sensors while reaching inferences at different scales. One open problem is the design of network topology including sensor placement for this specific application. The hierarchical network should be fault-tolerant so that failure of some nodes will not cripple the system.

Sensor suites at higher levels in the hierarchy have more responsibility in that they execute more elaborate communication protocols, monitor whether various sensors function correctly, and conduct decision-making while taking into account the information provided by other sensors. Each intelligent sensor suite consisting of multiple sensors in the overall sensor network is capable of making some decisions based on its own inputs. These decisions are passed on to higher-level nodes in the control hierarchy for information assimilation or information fusion. These decision-making problems can be formulated as hypothesis testing problems in a distributed framework. Many fundamental results on distributed detection are available in the literature [11]. But many open issues in distributed detection need further investigation, including decision making that includes domain-specific knowledge, detection and classification under uncertainty, and application of learning and data mining while making inferences.

4.2 Control and Response Planning

After drawing inferences regarding the current state of the building at various desired scales, control and response plans may need to be initiated. The control of the hierarchical IEQ system involves operating at many levels of focus, specificity, accuracy, and time constants. Automating such a multi-level control system is necessary in order to generate the automatic responses within short time frames dictated by emergency situations as well as for the more routine situations. The most logical architecture for this application is a multi-level control system that follows the structure of the multi-level intelligent system described earlier. In this case, lower level controls will deal with systems such as local HVAC, elevators, lighting, etc. The use of intelligent models in their controls will improve their performance, and also make their parameters adaptable to changing modes of operation, as monitored and activated by higher supervisory levels. Higher-level controls will view most events in a discrete domain, and will handle routine tasks such as coordination of local and building HVAC subsystems, power management, as well as unusual tasks such as responses to emergency situations.

In this application, "model-based" control will result in a control system that is pro-active rather than reactive and is desirable. Open issues in this area include development of intelligent control techniques at all levels as well as the interactions between levels. It is anticipated that the biggest advances in this area can be made in low-level process control, by utilizing analytic, neural network, fuzzy logic, and hybrid models as a part of the control system.

The other issue is that of dynamic response planning especially in emergency situations. The sensor network augmented with information fusion techniques will provide information and inferences to quickly and reliably detect and identify toxins, fire, and other dangerous environmental conditions. Based on the type and location of the danger, evacuation and mitigation plans can be more effectively implemented. Algorithms that automatically convert the information available on the network into plans of action are required. They will reduce response time and provide emergency personnel with critical knowledge before arriving to the emergency scene, saving the lives of more victims and emergency response team members.

The evacuation plan will identify all safe evacuation routes and communicate these routes to evacuees. These routes will minimize the evacuees' exposure to heat, smoke or invisible gases. These routes can be lit up with emergency lights and sound alarms to guide people out of the building safely. These routes can be used to minimize the probability that anyone becomes trapped. Control actions can also be sensitive to the evacuation plan, e.g., minimizing contaminant levels in areas of the building where humans are expected to pass through in the near future.

5. Examples

In this section, we present two examples of some ongoing work in the area of sensor networking for monitoring and control of IEQ in buildings. The first briefly discusses the work at the University of California at Berkeley for the routine situation. The second addresses evacuation planning in emergency situations and is based on ongoing work at Syracuse University.

Multisensor Control of Building IEQ Traditional systems for maintaining IEQ in buildings employ a single sensor to monitor the IEQ, e.g., temperature, of several rooms and then control it by means of a single actuator. As indicated earlier, sensor networking can provide information at a much finer resolution that can be used for more refined control. A number of research efforts are underway where the feasibility of such systems is being examined primarily to improve human comfort and for reduced energy consumption. For example, a single actuator control system based on information from multiple sensors has been studied by simulation [10].

A number of *ad hoc* information fusion strategies have been employed and researchers have studied optimization approaches to maximize comfort or to optimize energy consumption under comfort constraints. The study shows that the energy-optimal strategy reduced energy consumption by 17% while improving the comfort metric (Predicted Percentage Dissatisfied, PPD) from 30% to 24%. The comfort-optimal strategy reduced energy consumption by 4% while reducing PPD from 30% to 20%. These results are quite convincing in that they demonstrate the feasibility of significant improvement when sensor networking along with suitable information processing and control algorithms are used. This field is in its infancy and many research directions need to be pursued to enable the realization of the envisioned system.

Developing Evacuation Plans in Response to CBA/IEQ Emergencies Emergency evacuation plans have been considered in the past, largely with focus on fire emergencies [9, 2, 5]. However, the nature of biological and chemical contaminant flow in indoor environments differs substantially from the manner in which fire and smoke spread through a building; evacuation requirements also differ (e.g., safe evacuation time periods are shorter), as do remediation measures. This motivates our ongoing study of the formulation of evacuation plans tailored to biological or chemical attacks. The main idea in this work [13] is the coordinated application of building layout information, dynamically changing sensor outputs, physics based flow models, and optimization algorithms, in order to achieve evacuation path planning. The challenge is to employ all these factors that are highly dynamic and derive solutions that satisfy multiple objectives. The primary goal is to minimize the total exposure of occupants over a certain time interval under several constraints based on physics and building layout. An iterative procedure that includes a multi-zone air and contaminant flow network model, and applies an evolutionary algorithm for optimization, is employed to iteratively change control strategies and determine the evacuation path. A variety of scenarios were considered that included change of CBA/IEQ attack source, control plans, objectives, and optimization algorithms. This work has shown that a sophisticated sensor-based pervasive computing system can be valuable for ensuring safety of occupants in an emergency.

6. Concluding Remarks

We have examined various aspects of developing intelligent sensor networks in the context of built environments. The development of such a network is a challenging task, requiring careful choices and balancing tradeoffs between considerations such as cost and the ability to detect and respond to potentially life-threatening emergencies. Issues to be considered include the choice of sensors, network architectures, network communication strategies, and also data analysis methodologies that include

control and response plan formulation using iterative methods and optimization algorithms. Examples were given to illustrate the application of these approaches to apply sensor networks and analyze their data to maximize human comfort and safety. Substantial future research issues remain, and significant advances are expected to occur rapidly in improving indoor environments with networked sensors.

References

[1] Brambley, M.R., D.P. Chassin, K. Gowri, B. Kammers and D. J. Branson. "DDC and the Web." ASHRAE Journal, Volume 42, No. 12, pp.38-50, 2000

[2] Chalmet, L. G., Francis, R. L., and Saunders, P. B., "Network Models For Building Evacuation," Management Science, Vol. 28, No. 1, pp. 86-105, January 1982.

[3] Dexter, A.L. "Intelligent buildings: Fact or Fiction?" International Journal of HVAC&R Research, ASHRAE, vol.2, no.2, April 1996, 105-106. ISSN 1078-9669

[4] ETSI, Broadband Radio Access Networks (BRAN): HIPERLAN Type 2 Technical Specification Part 1 — Physical Layer, Oct. 1999.

[5] Galea, E. R., M. Owen and P. J. Lawrence, "Emergency Egress from Large Buildings under Fire Conditions Simulated Using the EXODUS Evacuation Model." in Proceedings of the 7th Intern Fire Science and Engineering Conference: Interflam'96, pp 711-720, St John's College, Cambridge, England, March 1996, compiled by C Franks and S Grayson, Published by Interscience Communications Ltd, London, UK, 1996. ISBN 0 9516320 9 4.

[6] Garg, V. and Wilkes, J., "Wireless and Personal Communications Systems," Prentice Hall, Upper Saddle River, NJ, 1996.

[7] Heinzelman, Wendi Rabiner, Anantha P. Chandrakasan, and Hari Balakrishnan, "Energy- Efficient Communication Protocol for Wireless Microsenso Networks," Proceedings of the 33rd Hawaii International Conference on System Sciences, pp. 1-10, 2000.

[8] IEEE, Wireless LAN Medium Access Control (MAC) and Physical Layer (PHY) Specifications, IEEE Standard 802.11a, 1999.

[9] Kisko, T. M., and R.L. Francis, "EVACNET+: A Computer Program to Determine Optimal Building Evacuation Plans," Fire Safety Journal, 9:211-222, 1985.

[10] Lin, C., C. C. Federspiel, and D. M. Auslander, "Multi-Sensor Single-Actuator Control of HVAC Systems," International Conference for Enhanced Building Operations, Austin, TX, October 14-18, 2002.

[11] Varshney, P.K., Distributed Detection and Data Fusion, Springer-Verlag, 1997.

[12] Zhang, J., Kulasekere, E.C.; Premaratne, K.; and Bauer, P.H. "Resource management of task oriented distributed sensor networks," Proceedings of the IEEE International Symposium on Circuits and Systems, pp. 513-516, Sydney, Australia, May 6-9 2001.

[13] J.S. Zhang, C.K. Mohan, P. Varshney, C. Isik, K. Mehrotra, S. Wang, Z. Gao, and R. Rajagopalan, "Intelligent Control of Building Environmental Systems for Optimal Evacuation Planning", Proceedings of International Conference on Indoor Air Quality Problems and Engineering Solutions, July 21-23, 2003, Research Triangle Park, NC.

Chapter 12

A DISTRIBUTED TRANSMITTER FOR THE SENSOR REACHBACK PROBLEM BASED ON RADAR SIGNALS*

Lav R. Varshney and Sergio D. Servetto
School of Electrical and Computer Engineering
Cornell University, Ithaca, NY 14853
{lrv2@,servetto@ece.}cornell.edu

Abstract We consider the problem of reachback communication in sensor networks. In this problem, a large number of sensors are deployed on a field, to measure the state of some physical process that unfolds over the field and to then cooperatively send this information back to a distant receiver for further processing. We formulate the problem as a multiple-input, single-output (MISO) system, and develop a time-division scheme based on transmission of simulated radar echoes. Information is encoded in the spatial electromagnetic reflectivity function of virtual point reflectors, and decoded with a conventional range radar receiver. Transmitter diversity and the use of pulse compression radar waveforms are exploited for both increased reliability and increased data rate. Information theoretic and simulation-based performance characterizations are also presented.

1. Introduction

1.1 The Problem of Remote Sensing

The problem of surveilling a scene from a distance arises in numerous contexts including disaster recovery, tactical battlefield assessment, and environmental monitoring, where proximate surveillance is not practical. Traditional methods of remote sensing have included imaging systems such as optical, multispectral, and synthetic aperture radar (SAR) [11]. Despite the prominent successes of these systems, the base sensing modalities suffer from shortcom-

*Work supported by the National Science Foundation, under awards CCR-0238271 (CAREER), CCR-0330059, and ANR-0325556. This manuscript is a summary of the first author's Electrical and Computer Engineering Honors Project Report.

ings. When operating at great distances, these sensing modalities achieve low resolution. Furthermore, the types of physical processes that can be sensed at a distance are limited to phenomena that propagate over large distances, such as electromagnetic waves. Other processes such as chemical, biological, or barometrical processes cannot be easily sensed.

Recently, wireless sensor networks made up of large numbers of small, inexpensive, but unreliable nodes with sensation, computation, and communication capabilities have been proposed for many surveillance applications. Ad situ deployment of large numbers of sensors offers solutions to both the limited resolution and limited sensing modality problems. Although not strictly remote sensing, since the data is sensed proximally, sensor network solutions may be able to retain distal advantages if the sensed data can be transmitted to a remote receiver. We refer to this transmission of sensed information from the sensor network to a distant receiver as sensor reachback communication.

If each of the sensor nodes were equipped with a powerful, reliable transmitter, capable of individually reaching a distant receiver, then a proximal sensing, distal data retrieval solution could be easily achieved. Due to severe energy constraints and inherent unreliability in most sensor networks, individual nodes do not have enough resources to independently reach a distant receiver, hence some form of cooperation among nodes is required for reliable reachback communication. In this paper we consider practical methods for implementing the uplink of one such sensor-assisted remote sensing system.

1.2 Sensor-Based Radar Imaging

Consider a remote sensing system design in which the transmitter, instead of illuminating a target with microwave radiation, spreads a large number of sensors over the target. These sensors collect some local measurements (the equivalent of measuring reflections by local scatterers in standard radar), which then need to be sent back to the radar transmitter. This scenario is illustrated in Fig. 12.1.

In this new radar architecture, sensors can be thought of as "programmable" scatterers. This is very important, since sensors can provide significantly more information than that provided by natural scatterers. Despite many advantages of microwave radar (such as being able to penetrate cloud cover and operate at night), a well known disadvantage of these systems is that it is difficult for them to achieve the resolution of *optical* imaging techniques [12]. And one of the most challenging aspects of research in this field is precisely the development of sophisticated signal processing algorithms, capable of extracting information from those coarse resolution measurements. The problem however is that even optimal processing can only extract as much information as is contained in the original signal—but not more, a consequence of the *Data Processing Inequal-*

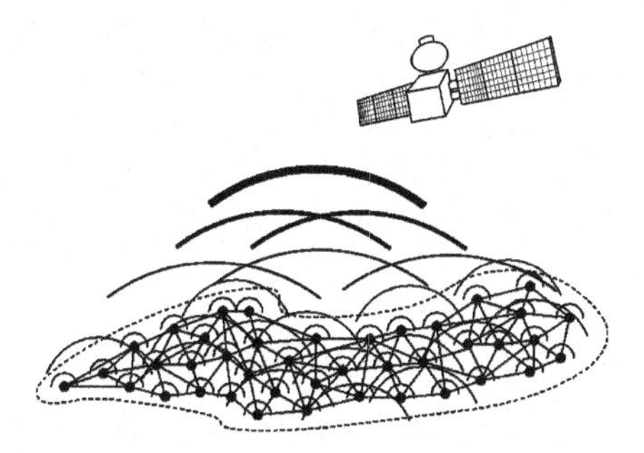

Figure 12.1. A large number of sensors are deployed over a target area. After collecting the data of interest, the sensors must *reach back* and transmit this information to a single receiver (e.g. an overflying plane, or a satellite) for further processing.

ity [6, pg. 32] in information theory. The main challenge to increasing the data rate achievable for the remote sensing systems under consideration therefore seems to lie in generating signals that contain more information.

There are numerous ways that one can design a distributed transmitter in a single-user multiple antenna system including various beamforming and space-time coding techniques. Our scheme is novel in that it is compatible with a prime remote sensing modality, SAR; in fact the signal generation of the distributed transmitter is matched to a standard radar receiver, allowing the existing radar receiver hardware, signal processing, and target recognition infrastructure to be leveraged for reachback communication. Since our scheme is based on SAR, individual sensor nodes transmit simulated radar return echoes, creating a simulated field of virtual point scatterers. The message to be sent is encoded in the spatial electromagnetic reflectivity function of these programmable virtual point scatterers and is received and decoded using a standard range radar receiver. In SAR signal processing, the two parameters used for spatial target discrimination are time and frequency shifts to determine range and azimuth respectively. These basic radar parameters lead directly to the possibility of a time- and frequency-division encoding scheme that can be processed with a standard radar receiver.

In multiple-input, single-output (MISO) systems, two types of gains can be provided, integration gains and multiplexing gains. Integration gains result in increased reliability by having multiple elements transmit signals carrying the same information. Multiplexing gains result in increased data rate by having elements transmit signals carrying different information at once. Although both types of gains can be achieved simultaneously, there is a fundamental tradeoff between the two. To achieve integration gains, we create cohorts of nodes that transmit the same signal in chorus. We use a single orthogonal direction, time, for separating various cohort signals, thus compromising multiplexing for integration. We are thus able to achieve reliable communication with very unreliable transmitter nodes. By using only time orthogonality, the system inherently exploits only radar range as a discriminating factor between transmitted symbols, and therefore requires only a simple range radar receiver. The communications scheme has minimal complexity at the receiver and particularly at the transmitter nodes.

1.3 Main Contributions and Paper Organization

The main original contributions presented in this work are the complete design, thorough performance analysis via simulations, and preliminary results on an information theoretic analysis of capacity for one specific class of low complexity sensor-assisted remote sensing system.

The rest of this paper is organized as follows. In Section 2 we discuss the system architecture of our sensor-assisted remote sensing system, and highlight all the relevant parameters. In Section 3 we report on the result of a number of numerical simulations, in which we seek to understand the effect of the different system parameters on the data rates achievable in our uplink. Some initial results on an information-theoretic characterization of the capacity of systems operating under the constraints described in Section 2 are presented in Section 4. Related work is discussed in Section 5. Concluding remarks are presented in Section 6.

2. System Architecture

2.1 General Conception of the Sensor Reachback Communication System

Following the classical Shannon pentapartite division of a general communication system, the information source for sensor reachback communication is defined as a field of sensors that measure some physical phenomenon. The data samples detected by the individual sensor nodes are correlated, since physical processes display high spatial correlation and continuity. Through an underlying internode communications infrastructure and information dissemination protocol in the sensor network, an estimate of the entire field of measurements can be formed at every sensor [15]. The entire field of measurements, encoded into a single network-wide discrete alphabet, forms the message that is to be transmitted.

The transmitter is composed of individual sensors (equipped with radio transmitters), that work cooperatively to send a signal to a central, distant location. Collectively, the individual sensors form a distributed radio transmitter antenna array, capable of generating an aggregate waveform. The design of the distributed transmission protocol to produce the aggregate waveform forms the bulk of this study. The general concept of this reachback communication protocol is for the individual sensor nodes to simulate radar return echoes from point scatterers. The message to be sent is encoded in the spatial electromagnetic reflectivity function of the "programmable scatterers."

For wireless sensor reachback, transmission is achieved by direct path radio wave propagation through air, thus we assume an additive white Gaussian noise (AWGN) channel. Since a large number of sensors wish to communicate with a common receiver over a common channel, one might consider this as a multiple access channel. Alternatively, the system may be considered to have a single transmitter and a single receiver, where the multiple nodes work together to form a single signal.

The receiver determines the message sent by the transmitter, decoding the message encoded in the aggregate waveform and delivers it to the destination, where it is used. The receiver to be used is a standard high range resolution radar receiver, with standard radar signal processing. This results in a pulse amplitude modulated signal, which can be decoded using standard pulse amplitude modulation decoders. If the system is used for binary signaling, on-off keying as a special case of pulse amplitude modulation would readily apply.

A schematic diagram of the reachback system is given in Fig. 12.2.

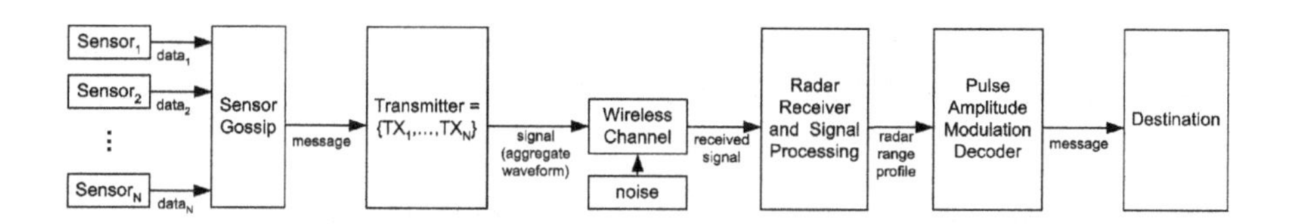

Figure 12.2. The sensor reachback communication system in the Shannon framework.

2.2 Geometry and Topology

In this study we consider a square lattice network topology, however the results may be generalized to networks with randomly distributed nodes. By considering this square lattice configuration, we inherently assume that nodes are able to determine their spatial location. Fig. 12.3 shows the network topology. For simplicity in exposition, we assume that the number of nodes is the square of an odd integer; hence there is a node located at the center of the network. For optimal performance, the receiver is positioned along one of the central axes of the network, as shown in Fig. 12.4. The nodes in the network

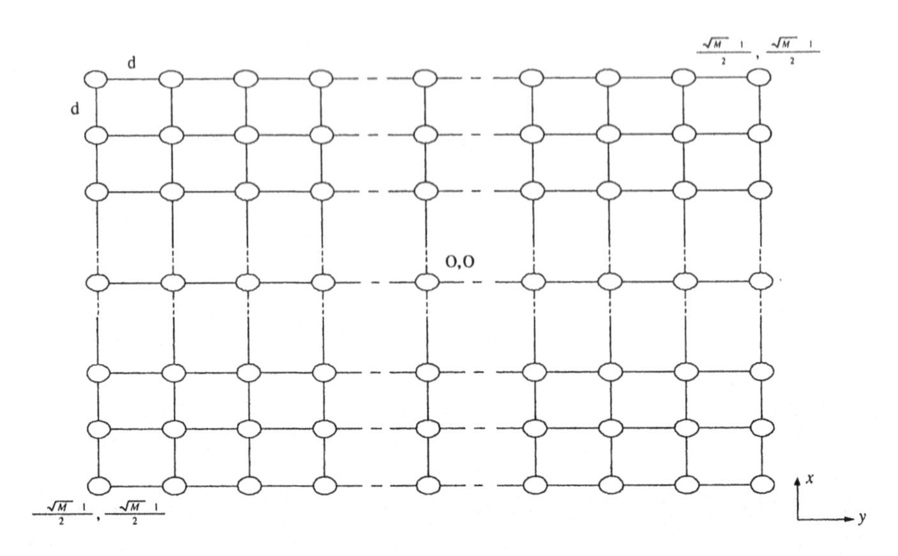

Figure 12.3. Network topology. There are M nodes in the network, and the fixed distance between nodes in the lattice is d. The discrete node locations are indexed such that $\frac{x}{d} = \frac{-(\sqrt{M}-1)}{2}, ..., -1, 0, 1, ..., \frac{(\sqrt{M}-1)}{2}$ and $\frac{y}{d} = \frac{-(\sqrt{M}-1)}{2}, ..., -1, 0, 1, ..., \frac{(\sqrt{M}-1)}{2}$.

are allocated to different cohorts, based on their location along the central axis that is chosen. For randomly distributed nodes, division into cohorts would be done in the same manner.

2.3 Radar Range Processing and Range Resolution

The problem of standard radar range processing is one of estimating the time between signal transmission and reflected echo reception. For the one-way communications scheme that uses time orthogonality, we are simply interested in separating the different cohort signals; therefore simulated echo arrival time is sufficient. The range resolution, ΔR, is the minimum discriminable distance between two targets in standard radar, and the width of a range cell for the distributed targets measured in SAR. The range resolution determines the maximal multiplexing that is possible and limits the data rate for a given system. The range resolution of a simple pulsed transmitter for one-way transmission is cT, where c is the propagation speed and T is the pulse width. The range resolution can be made arbitrarily small by reducing the pulse width, but under transmitter peak power constraints, this leads to a reduction in pulse power, and therefore reliability.

By using specific phase modulation techniques in the pulse waveform, the range resolution advantages of short pulses can be obtained with long pulses,

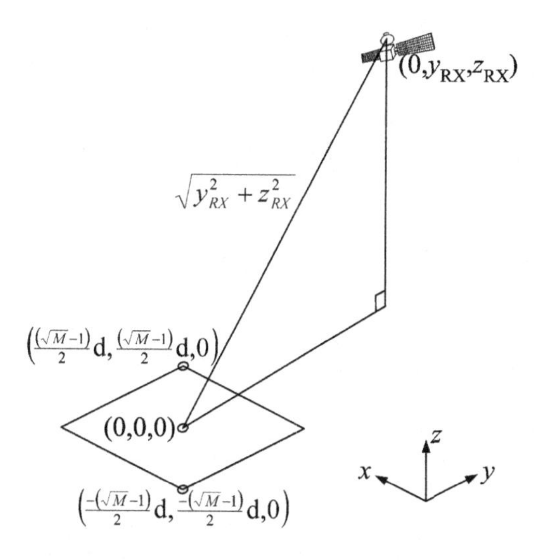

Figure 12.4. Transmitter and receiver geometry. The receiver is situated along the y-axis, therefore division into cohorts would be performed so that members of a single cohort have nearly the same y-coordinate. The number of cohorts, C, is such that $1 \leq C \leq \sqrt{M}$ (at least one range line per cohort).

while staying within peak power limits. Through the process of pulse compression, which is accomplished through the use of a matched filter, the modulated long pulse with increased bandwidth B can be compressed in the receiver to a compressed pulse with a main lobe width approximately equal to $1/B$. There are numerous classes of waveforms that are suitable for pulse compression [4]. Complete characterizations of waveforms are given by the radar ambiguity function [16], a two-dimensional correlation function with respect to time- and frequency-shifts and represents the pulse compression matched filter output. For our application, the waveforms may be considered to be simply defined by the pulse compression ratio, R_{pc}. The R_{pc} is defined as the ratio of the uncompressed pulse width T to the compressed pulse width $\tau \approx 1/B$. Thus the pulse compression ratio, R_{pc}, is approximately equal to the time- bandwidth product of the waveform. The range resolution of a pulse compression waveform is given by

$$\Delta R = c\tau = \frac{cT}{R_{pc}} = \frac{(\Delta R)_{uncompressed}}{R_{pc}} \approx \frac{c}{B}. \tag{12.1}$$

As seen, R_{pc}-fold finer range resolution is possible through phase coding. An additional effect of pulse compression processing is that the energy of the entire long pulse is compressed into the short pulse, an effect known as pulse compression gain.

2.4 Generating Range Profiles, Aggregate Waveforms

All transmitter nodes in the network transmit a fixed pulse compression waveform. For the purposes of encoding information, the transmitters in the sensor network are divided into C cohorts, as described in Section 2.2. Through the information dissemination protocol, all members of a cohort know the symbol that the cohort is to transmit at any given time. The symbols are encoded in the amplitude of the transmitted pulse and all members transmit the symbol simultaneously. The contributions of all cohort members are integrated with some integration efficiency, producing a symbol received at the receiver. Range processing is performed on the received signal to determine the contributions of the different cohorts on the range profile. If all cohorts transmitted simultaneously, difference in propagation delay would be the only discriminating characteristic as in standard radar range processing, and would require very fine range discrimination and receiver positioning. In the context of this problem, where transmitter nodes are located over a small area, and the receiver position is unknown by the transmitters, artificial delays are introduced to reduce the need for fine range resolution and receiver positioning.

Each cohort is assigned an artificial delay at which time all members of the cohort transmit the signal. The artificial delay assignment for the kth cohort is

$$\alpha_k = k\frac{t_A}{C}, k = 0, 1, ..., C-1, \tag{12.2}$$

where $ct_A = R_A$ is the desired total artificial range separation between the most distant cohorts. The introduction of an artificial delay serves to increase apparent range separation between transmitters, thereby allowing coarser range resolution, which is often required by radar receiver hardware. If the most distant cohorts have small receiver line of sight range separation, R_P, as compared to R_A, then the propagation delays can be treated as negligible. This is a desired condition for robustness to receiver location. Since cohort signals are discriminated by artificial delay, rather than propagation delay, the propagation delay can be considered a source of noise. The differences of propagation delay within the cohorts, due to the spatial distribution of nodes, cause a broadening of the symbol peak and difference of propagation delay among the different cohorts may cause the equal spacing between peaks to become unequal.

From simple geometry, the distance between a node in the network and the receiver may be defined as

$$d\left(x_{TX}, y_{TX}\right) = \sqrt{\left(x_{TX}\right)^2 + \left(y_{RX} - y_{TX}\right)^2 + \left(z_{RX}\right)^2}, \tag{12.3}$$

for the geometry considered in Fig. 12.4. When propagation effects are considered, rather than appearing at the artificial delay location as desired, the peak of the kth cohort symbol will occur at

$$\beta_k = \alpha_k + \tfrac{1}{c} d\left(\bar{x}_k, \bar{y}_k\right), \tag{12.4}$$

where the argument of the distance function is the centroid of the cohort, calculated with a distance density measure. Propagation effects also cause the narrow peaks of the ideal pulse compressed range profile to be broadened. This effect is like convolving the narrow peak with a filter that has time support of length

$$t_{B,k} = \tfrac{1}{c}\left[\max_{i \in C_k} d\left(x_i, y_i\right) - \min_{i \in C_k} d\left(x_i, y_i\right)\right], \tag{12.5}$$

and shape like the distribution of the $d(x_i, y_i)$ in cohort k.

2.5 Nodes, Cohorts, Distributed Transmitter, and Receiver

The individual transmitter nodes can be defined by a small number of parameters. The pulse compression waveform is selected for its pulse compression ratio, and is fixed for all transmitter nodes and for all time. This pulsed waveform, of finite time duration T, is denoted $s\left(t\right)$. The artificial delay and the symbol amplitude for transmission are determined by membership in the cohort. The symbol amplitude changes for each new message, whereas the relative artificial delay is constant for all time. One can consider propagation delay as a property of the node rather than the channel; thence propagation delay is incorporated into the node itself, and so the output of each node i in cohort k before amplitude scaling is given by

$$AX\left[n\right] s\left(t - k\tfrac{\Delta t_A}{C} - \tfrac{1}{c} d\left(x_i, y_i\right)\right) = AX\left[n\right] s\left(t - \alpha_k - t_{P,i}\right), \tag{12.6}$$

where $t_{P,i}$ is the propagation delay for the node, and $X\left[n\right]$ is a discrete-time pulse amplitude scaling factor, drawn from a Q-ary alphabet, with a suitable normalization constant A to ensure that the average output energy of the system is fixed regardless of the number of nodes. The superposition of these individual node signals is the signal generated by a cohort is

$$AX\left[n\right] \sum_{i=0}^{N_k - 1} s\left(t - \alpha_k - t_{P,i}\right), \tag{12.7}$$

for a cohort with N_k nodes.

Since each of the cohorts begin transmission at a different artificial delay time, the transmitter may be modeled as a time-division scheme even though the pulses may overlap. Thus the transmitter may be given schematically as in Fig. 12.5 using a commutator. The figure also shows the channel and the receiver. The receiver to be used is a standard range radar receiver. It receives the transmitted signal, performs range compression, and then uses a sampler and maximum likelihood decoder to convert the continuous-time, continuous-valued output into a discrete-valued output sequence so as to determine the encoded message. The received signal is given by the superposition of the signals generated by all nodes that are members of all cohorts and noise:

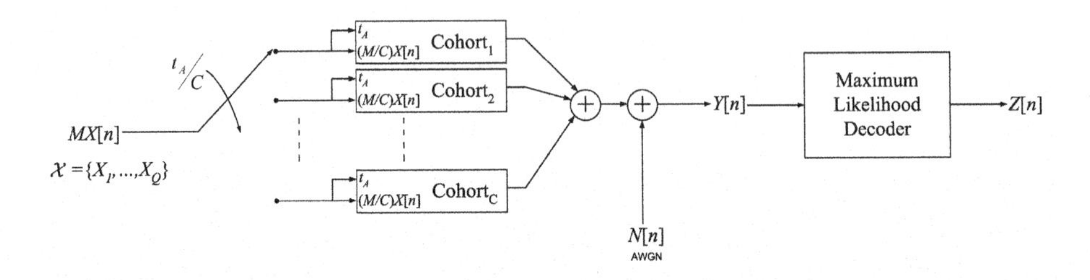

Figure 12.5. Entire system without regard to sampling. Note that pulse compression is not shown explicitly, but is included in the maximum likelihood decoder. The noise power of the additive white Gaussian noise is fixed. The symbols X are drawn from a Q-ary alphabet such that the expected value of the output $S = \mathrm{E}\{X\}$ is inversely proportional to M, the number of nodes. Thus the total average output of the network is fixed regardless of size. Thus the signal to noise ratio is fixed regardless of size. The switching period of the commutator switch t_A/C is determined by the total artificial time separation among all cohorts, t_A, and the number of cohorts. This value will be related to the range resolution of the waveform chosen.

The received signal is pulse compressed using matched filtering to generate

$$m(t) = \int_{-\infty}^{\infty} y(\xi)s^*(\xi - t)\, d\xi, \tag{12.9}$$

which is then detected with a maximum likelihood pulse amplitude modulation detector which quantizes to determine the estimate of the input symbol.

2.6 Data Rate

Three factors, the number of cohorts, the number of possible symbols that each cohort can send, and the time between successive transmissions determine the data rate of the transmission protocol. If each of the M cohorts select symbols from a Q-ary alphabet and transmit a symbol every T_{PRI} seconds, then the transmission rate will be $\frac{M \log_2(Q)}{T_{PRI}}$ bits per second. As the number of cohorts increases, the membership sizes of the cohorts decrease, thereby reducing the number of signals that are superposed and the symbol power. The range resolution acts to bound the number of symbols that can be sent in a given period of time, since symbols that are closer in range than the range resolution cannot be discriminated. The size of the alphabet from which symbols are chosen is determined by the noise in the system to a large degree. The pulse compression ratio is fundamental in determining both of these parameters; it is inversely related to range resolution, and is related to the pulse compression gain by a square-root relationship. Increasing the pulse compression ratio allows better range resolution allowing larger M and a greater signal to noise ratio, allowing the possibility of a larger Q. Every time a set of pulses is transmitted, a packet of M symbols is sent. The third factor in the rate is the number of packets that are sent every second. In pulsed radar, this interpacket time is referred to as the pulse repetition interval (PRI). Clearly, the pulses may not overlap, so the PRI must be greater than the pulse width. When low duty cycles are used, the PRI will be much greater than the pulse width. In the reachback problem there is no explicit restriction on the PRI, except that which may be imposed by energy constraints or receiver design.

3. Simulation-Based Characterization

In order to characterize the proposed sensor reachback communication scheme based on radar signals, a simulation was developed. The signaling scheme chosen was an on-off keying (OOK) pulse amplitude modulated Barker binary phase-coded waveform. Rather than choose a carrier frequency, the complex envelope was used; the complex envelope is a baseband waveform that almost completely describes the waveform. The output signal sampling times and the maximum likelihood decoder threshold were determined *a priori* from system parameters, and did not involve any adaptive processing for impairment mitigation, thus may not achieve the best possible performance in the presence of various impairments.

We performed simulations to measure the effects of various system parameters on performance. The fundamental limits of reliable communication for any communications scheme are imposed by noise. By varying the signal to noise ratio (SNR) for the additive white Gaussian channel, we determine the symbol error rate (SER) for a given system configuration. Fig. 12.6 shows the relationship between output SNR and SER determined by performing Monte Carlo simulations for various noise levels with a Bernoulli($\frac{1}{2}$) source, and two possible system configurations – the difference is in the increased number of cohorts, which causes an increase in the time transmission rate.

Although signal strength decreases inversely as the square of distance, we do not explicitly take this into account, as it is captured in the SNR. Similarly,

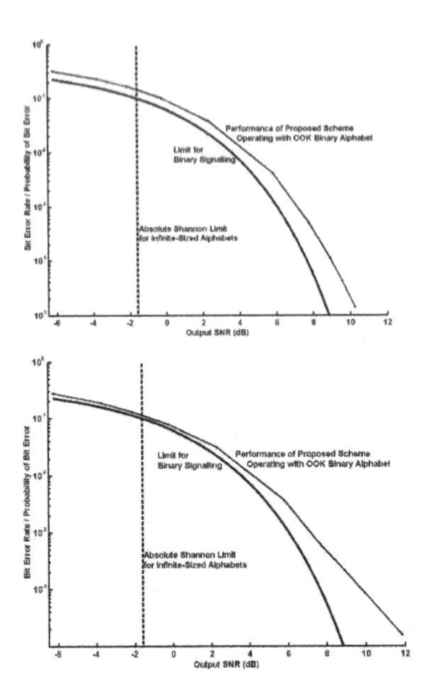

Figure 12.6. Bit error rate as a function of output signal to noise ratio. A Barker 13 phase coded waveform with pulse compression ratio of 13 (22.3dB) was used in both figures. For the figure on the left, the the transmission rate would be 2.1×10^6 bits per second for a 100% duty cycle, the number of nodes in the network was 3969, and there were 7 cohorts. For the figure on the right, the transmission rate would be 6.3×10^6 bits per second for a 100% duty cycle, and there were 21 cohorts. The physical area occupied by the network was 200m$\times$200m, and the receiver was situated 10km behind and 50km above the center of the network. For comparison, a binary signaling bound and the fundamental bound for arbitrary signaling alphabets are shown.

various antenna gains, receiver noise figures and system losses are not explicitly considered. Note that the SNR given is the output SNR, calculated after performing pulse compression of the received signals by matched filtering; hence the pulse compression gain, which is equal to the pulse compression ratio, has been included in the signal power. We use OOK Barker waveforms, which have pulse compression gains equal to the length of the phase code. Further note that the signal power is measured as the average output of all nodes in the entire network. For comparison, the probability of bit error as a function of output SNR for orthogonal binary signaling over an AWGN channel is also given. This limit is the probability of error for an AWGN channel used as a binary symmetric channel, with pulse compression gain taken into account, and is given by the Gaussian cumulative distribution function. Also shown is the fundamental Shannon limit of performance for any signaling alphabet over an AWGN channel, a vertical line at -1.59 dB. It is known that as the size of the of the orthogonal signaling alphabet approaches infinity, the probability of error curve approaches the vertical line at -1.59 dB. Thus orthogonal signaling alphabets are asymptotically optimal for an AWGN channel [3]. The causes for suboptimal performance are mainly interference between different cohort channels, resulting from overlapping sidelobes in the pulse compressed domain, and the fact that the transmitter was distributed and did not achieve perfect integration gain due to different propagation delays.

Another fundamental parameter in the system is the number of cohorts into which the nodes are divided. Since the total number of nodes in the network is fixed, increasing the number of cohorts reduces the number of nodes in each cohort, causing a classic multiplexing rate gain-integration reliability gain tradeoff. Since our system model has an additive white Gaussian noise channel, there would be no diversity gains as is found in systems with fading channels. Fig. 12.7 shows the symbol error rate as a function of output SNR for various numbers of cohorts. Remember that as the number of cohorts increases, the transmission rate also increases. As can be seen in the figure, there are two regimes of performance, one for small numbers of cohorts and another for large numbers of cohorts. Although the ordinal performance within these two groups does not appear to be related to the actual number of cohorts, it is clear that there is a transition point. This transition point is closely related to the point when interchannel interference becomes very severe.

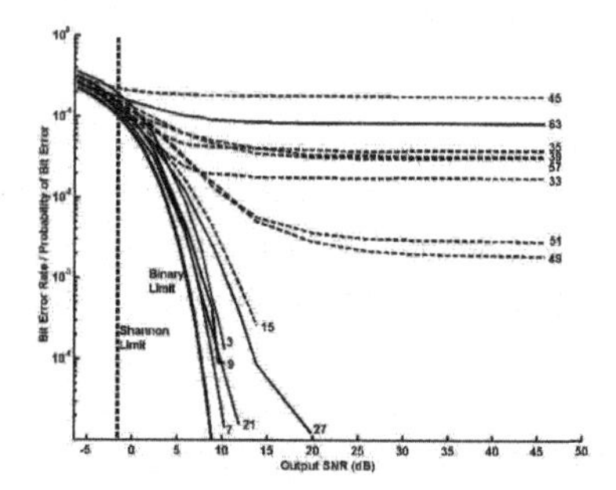

Figure 12.7. Bit error rate as a function of output signal to noise ratio for various numbers of cohorts. The number of cohorts is denoted to the right of the respective curve. The transmission rate would be 3×10^5 bits per second per cohort for a 100% duty cycle. A Barker 13 phase coded waveform with pulse compression ratio of 13 (22.3dB) was used. The number of nodes in the network was 3969. The physical area occupied by the network was 200m$\times$200m, and the receiver was situated 10km behind and 50km above the center of the network. For comparison, a binary signaling bound and the fundamental bound for arbitrary signaling alphabets are shown.

The pulse compression ratio is the main waveform parameter. Increased pulse compression ratio results in a narrower compressed pulse and therefore less main lobe interchannel interference. The pulse compression ratio is equal to the pulse compression gain, and therefore determines the output SNR for a given input SNR. By increasing the pulse compression ratio, the output SNR is increased. Thus it is expected that there should be no performance disadvantage in increasing the pulse compression ratio since reliability should increase with no decrease in rate. Increased bandwidth from increased pulse compression ratio may be a disadvantage in certain situations. Fig. 12.8 shows the performance of systems using different Barker coded waveforms with fixed input SNR. It is unclear why a waveform with pulse compression ratio of 5 performs better than ones with greater pulse compression ratios.

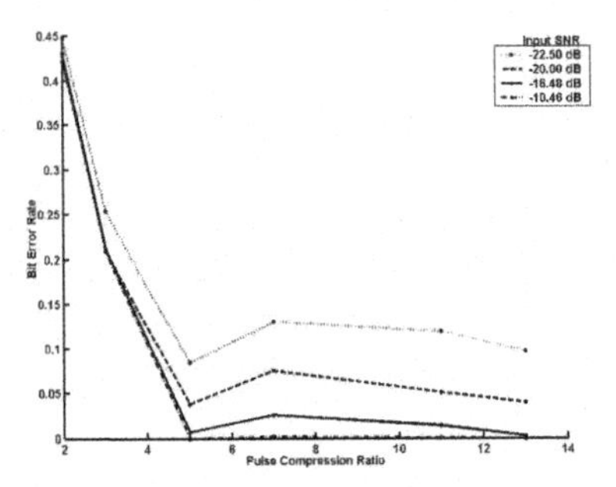

Figure 12.8. Bit error rate as a function of pulse compression ratio. The various curves show different input SNR values. Barker phase coded waveforms were used. The transmission rate would be 2.1×10^6 bits per second for a 100% duty cycle. The number of nodes in the network was 3969 and the number of cohorts was 7. The physical area occupied by the network was 200m$\times$200m, and the receiver was situated 10km behind and 50km above the center of the network.

We also simulate a departure from the ideal synchronization assumption by introducing timing jitter. Timing jitter for sensor networks has been modeled as a Brownian process [7], which for a fixed time is a mean-zero Gaussian random variable, independent for each node. We measure the reliability performance of the system as a function of the variance of the timing jitter. Reduced synchronization should reduce the integration gain of having multiple nodes transmit at the same time and increase interchannel interference by broadening the symbol. The decrease in performance is quite dramatic. Note however, that the decoder that was used was fixed *a priori* for no jitter, and this leads to additional degradation in performance, as opposed to a decoder which employs forms of estimation or adaptive processing. Fig. 12.9 shows results for the first system configuration used in Fig. 12.6.

One parameter of the system that has not been considered up to this point is the carrier frequency, since the baseband complex envelope signal has been

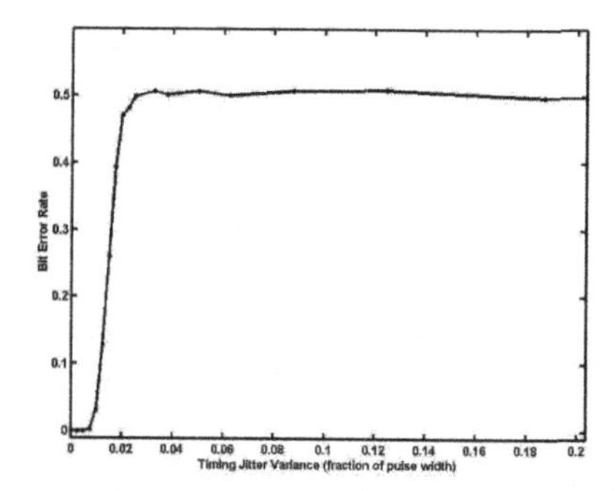

Figure 12.9. Bit error rate as a function of timing jitter variance. Output signal to noise ratio was 11.8213 dB. The transmission rate would be 2.1×10^6 bits per second for a 100% duty cycle. The pulse width was 1 μs. A Barker 13 phase coded waveform with pulse compression ratio of 13 (22.3dB) was used. The number of nodes in the network was 3969 and the number of cohorts was 7. The physical area occupied by the network was 200m$\times$200m, and the receiver was situated 10km behind and 50km above the center of the network.

used. Although this parameter may seem unimportant in theoretical discussions, it is of some importance for practical considerations. This parameter plays a role in determining the efficiency with which the waveforms of the several nodes in a cohort are integrated. In SAR, there is a well-known phenomenon called speckle, which is caused by localized destructive and constructive interference due to scatterers in a single ground cell differing in range by less than a wavelength of the carrier signal. In our scheme, a speckle phenomenon involving constructive and destructive interference may also occur depending on the carrier frequency, the physical dimensions of the system, and the physical arrangement of the several nodes in a cohort. The total number of nodes in a cohort, which had negligible impact for the baseband case, may also have some effect on the signal power that is received due to speckle.

Overall, the simulation-based characterization has shown the effect that some of the system parameters, including the signal to noise ratio, the number of cohorts, and the pulse compression ratio, have on performance. It has also been demonstrated that performance close to the orthogonal binary signaling alphabet limit may be achieved with the scheme. The importance of synchronization to the reachback communication scheme has also been demonstrated. We now move towards a strict information theoretic channel capacity characterization of the system.

4. Information Theoretic Characterization

In order to characterize the channel capacity of the proposed sensor reachback communication scheme, we first characterize the channel. We treat the channel as an additive MISO channel with additive white Gaussian noise. For the information theoretic characterization, we assume that a cohort can be modeled as a single transmitter antenna located at the centroid of the cohort, which is valid when the R_P/R_A ratio is much less than one. Under this assumption, reachback with radar signals is simply a time-division scheme, albeit before pulse compression there is actually overlap among the signals of the different cohorts. Time division generates orthogonal signals for each antenna element, thus the multiple-element antenna channel can be analyzed as a set of independent parallel channels.

The input to the antenna array is formed by time-multiplexing a scalar-coded symbol stream $X[n]$ across the cohorts. Symbols are dealt to the cohorts periodically, so that $X[n]$ is transmitted using cohort k when $n \equiv k(modC)$. A randomized time-division system would select a cohort randomly with uniform distribution, however the mutual informations associated with this scheme are identical to the deterministic counterpart [13]. The effects of the nodes that are members of the cohort contributing to the kth antenna are encapsulated in a single complex variable a_k, which represents the scaling and phase shifting resulting from aggregation. Similarly, the variable g_k is introduced to represent the effect of the channel encountered by signals emanating from cohort k. A simple evaluation of the radar range equation may be used to determine the signal power at a given distance from a cohort. The system can now be modeled as in Fig. 12.10.

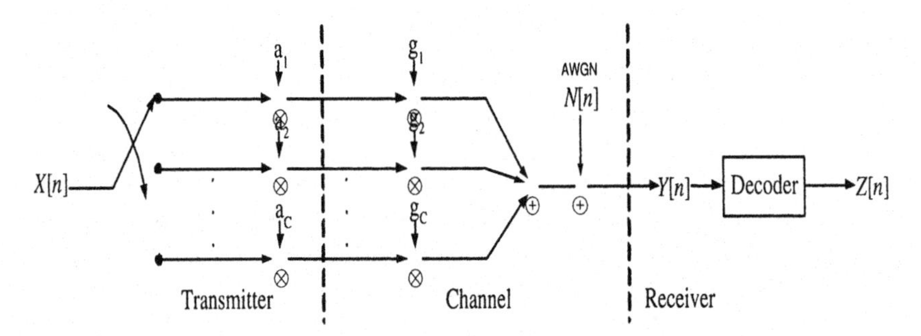

Figure 12.10. System model for information theoretic analysis.

The output for a single channel usage is

$$\left(a_{(n \bmod C)} g_{(n \bmod C)} X[n] + N[n]\right) \rightarrow \boxed{ML} \rightarrow Z[n] \in \{x_1, ..., x_Q\}$$
$$Y[n] \rightarrow \boxed{ML} \rightarrow Z[n] \in \{x_1, ..., x_Q\}$$

In order to determine the channel capacity of the reachback with radar signals scheme, we follow the methodology of [13], examining the mutual information between input and output over a long block of symbols, and find that the mutual information of the time-division channel is simply the average of the mutual informations achieved by each cohort.

If one considers the $Q = 2$ case, and the equiprobable input/output symbol alphabet $\{x_A, x_B\}$, the individual Gaussian cohort channel may be transformed into a binary symmetric channel through quantization of the output. If a maximum likelihood detector is used as the quantizer, the channel capacity of the individual cohort channel will be

$$I_k = 1 - H\left[\Phi\left(\frac{a_k g_k\left(\frac{1}{2}(x_A + x_B) - x_B\right)}{\sigma}\right)\right], \tag{12.10}$$

where $H(\cdot)$ is Shannon's binary entropy function, $\Phi(\cdot)$ is the error function, and σ is the noise variance. If OOK signaling is used, then the capacity formula reduces to

$$I_k = 1 - H\left[\Phi\left(\frac{a_k g_k x}{2\sigma}\right)\right], \tag{12.11}$$

where x is the symbol amplitude.

If one considers an input sequence $X[n]$ that is i.i.d. complex circularly symmetric Gaussian with energy $E|X[n]|^2 = x_s$ per symbol, it is the situation considered in [13], which has

$$I = \frac{1}{C}\sum_{k=1}^{C}\log_2\left(1 + \frac{|a_k|^2|g_k|^2 x_s}{\sigma/2}\right). \tag{12.12}$$

This can achieve the optimal performance for vector-coded antenna systems

$$I \leq I_{OPT} = \log_2\left(1 + \frac{\|\vec{a}\|^2\|\vec{g}\|^2 x_s}{C\sigma/2}\right) \tag{12.13}$$

when all cohort gains are equal, $|a_0 g_0| = |a_1 g_1| = \ldots = |a_{C-1} g_{C-1}|$.

This entire analysis has been done in discrete time, measuring mutual information per channel usage. In order to determine the continuous time capacity, the switching period of the commutator switch in Fig. 12.10, must be considered. The switching period is determined by the number of cohorts and the range of artificial delay assignments. Even though the capacity per channel usage is just the average of the individual channel capacities, a far greater continuous time rate is achievable with multiple cohorts. Overall CI bits of information can be transmitted reliably per PRI, and $\frac{CI}{T_{PRI}}$ bits per second.

5. Related Work

The work presented in this paper grows naturally out of an idea we outlined first in [2], of using radar imaging principles to implement the uplink of a sensor network. Other related work in our group deals with the development of time synchronization algorithms for the sensor array [8], and the development of a distributed FSK modulation scheme for the same application [9]. A different (but somewhat related) technique of energy accumulation for communication in a distributed setup has been considered in [10, 14].

There is one reference of which we became aware while our work was well under way, on which we need to comment in detail: work of a very similar nature was developed recently and independently by Ananthasubramaniam and Madhow [1]. Therefore, we feel it is only appropriate to highlight the similarities and differences among the their approach and ours.

Ananthasubramaniam and Madhow's scheme for sensor reachback communication operates without requiring a sensor network, in its classical sense, at all. No information dissemination occurs among sensor nodes, no distributed signal processing is performed, and no cooperation among nodes for reachback communication is required. Thus each node is wholly responsible for transmitting its message to the receiver, or "collector node." Our scheme becomes almost identical to their proposed scheme, if the artificial range separation is taken to be zero, the number of cohorts is expanded to the limiting case when the number of cohorts equals the number of nodes, and azimuthal processing, in addition to range processing as in true imaging SAR, is applied at the receiver.

Due to the lack of communication between sensor nodes in the scheme of [1], the underlying phenomenon being sensed is required to be in the form of discrete events that occur at random locations in the field, and the only message that may be transmitted is an activity map. Since spatial correlation is not reduced by distributed processing, there will generally be much inherent redundancy in the activity map, and this redundancy will in fact be the source of increased reliability in their scheme, very similar in spirit to the information theoretic work of Cover, El Gamal and Salehi [5]. In our scheme, however, an arbitrary stream of symbols may be sent, and standard forms of error correction codes may in fact be used for increased reliability in addition to the use of multiple nodes used to transmit the same signal synchronously.

In terms of synchronization, the scheme of [1] uses the collector node to transmit a beacon signal which is then actually echoed by the sensor nodes. In our scheme, no such explicit beacon signal is used. Rather, it is assumed that the nodes may be synchronized on the ground by some other means (e.g., [8]). Depending on the application under consideration, this may or may not be drawback. For example, for military applications, where the airborne receiver is the vital asset, it may not be so wise to require radio transmissions from the receiver, as has been the trend for military multistatic UAV-based surveillance systems. For other applications, such as in environmental monitoring, having the collector emit pulses does not pose problems.

Finally, the main conceptual difference between the work of [1] and ours is the way in which limitations in classical SAR systems enter into these systems. In classical SAR systems, the physical resolution for transmitting back sensed data in fundamentally limited by the physics of the system, typically greater than 10m for most systems. This source of limitation on the data rates attainable in the uplink does apply to the scheme of [1] as well, but not to ours. If the density of the sensor nodes is greater than the range/azimuth resolution of the imaging radar, then the signals from all nodes within a single range-Doppler cell will become superposed, in essence low-pass filtering the collected data. Furthermore, they will cause the speckle phenomenon, which will result from random constructive and destructive interference. So the scheme of [1] will not mitigate the two main drawbacks of SAR imaging, namely, low resolution and speckle noise, which was perhaps one of our main motivations in dealing with sensor-assisted remote sensing systems. Overall, having more densely placed sensors will actually not increase the amount of information that can be retrieved, since the ground is divided into discrete range-Doppler cells, and the cells' reflectivity coefficients can be considered as discrete-time, memoryless, continuous amplitude input data to a source encoder. This physical resolution problem was the primary reason that the concept of artificial range separation was introduced in our scheme. A similar artificial Doppler shift could also be introduced in our scheme if two-dimensional, range- and azimuth-discriminating, channel slicing was desired.

In summary. Both the systems described in [1] and ours are based on similar principles: have the sensor network emit signals resemblant of those dealt with by radar receivers. In the system of [1], the sensors act independently. This has the advantage of being implementable with what is perhaps the lowest possible complexity for any such approach, but is subject to the standard limitations of radar imaging systems. Our system, at the expense of extra complexity in the form of communication and coordination among sensor nodes forming a network, is able to operate in a regime that is *not* limited by the fundamental physical limitations of classical radar. Both schemes have merit, and which one is better suited for specific applications depends entirely on demands in terms of data rates achievable and complexity affordable by these applications.

6. Conclusions

In this work, a system architecture and signaling protocol for sensor reachback communication has been developed. The scheme has been designed in a "backwards compatible" way so as to allow the use of a standard radar receiver and signal processing, thus information is encoded in magnitude, with spatial discriminability allowing independent information to be transmitted from different range cells. Operating under the assumption that individual nodes will not have sufficient resources to signal a distant receiver, the scheme has incorporated synchronous signaling from an entire cohort of nodes that will be integrated together. Multiple cohorts are used to increase the transmission rate that is possible. Treating each cohort as a cohesive unit, the scheme can be described as a medium access control protocol that divides the channel into numerous independent channels. The division is partially based on inherent time orthogonality of transmitters located at different ranges due to propagation delays and is supplemented by the introduction of artificial range separation. Channel slicing is further supported by the use of pulse compression waveforms, which allow significantly time-overlapping pulses to be separated. Overlapping pulse compression sidelobes make the channels not truly independent.

Simulation results demonstrate that the scheme can achieve good performance, despite being designed under transmitter simplicity constraints and under the constraint of a specific type of receiver. Simulation has been performed only using binary OOK signaling using Barker coded waveforms. Further investigations into system performance by changing these parameters would provide greater insight into the potential of the scheme that has been proposed. In particular, it would be advisable to investigate linear frequency modulation waveforms that find great use in synthetic aperture radar. Furthermore, the use of adaptive pulse amplitude modulation decoders may be investigated for improved performance by mitigating impairments.

Future work into the mathematical characterization of the system may involve modeling node failure, timing jitter, and the speckle phenomenon as forms of channel fading. Cohort misassignment, stemming from poor localization, may also be characterized as a form of either noise or interference. Comparisons to traditional beamforming, space-time coding, and the other reachback communication schemes that have recently been developed may also follow.

References

[1] B. Ananthasubramaniam and U. Madhow. Virtual Radar Imaging for Sensor Networks. In *Proc. Int. Wkshp. Inform. Proc. Sensor Networks (IPSN)*, 2004.

[2] J. Barros and S. D. Servetto. On the Capacity of the Reachback Channel in Wireless Sensor Networks. In *Proc. IEEE Int. Workshop Multimedia Sig. Proc.*, US Virgin Islands, 2002. Invited paper to the special session on *Signal Processing for Wireless Networks*.

[3] R. E. Blahut. *Principles and Practice of Information Theory*. Addison-Wesley, 1987.

[4] C. E. Book and M. Bernfeld. *Radar Signals: An Introduction to Theory and Application*. Artech House, 1993.

[5] T. M. Cover, A. A. El Gamal, and M. Salehi. Multiple Access Channels with Arbitrarily Correlated Sources. *IEEE Trans. Inform. Theory*, IT-26(6):648–657, 1980.

[6] T. M. Cover and J. Thomas. *Elements of Information Theory*. John Wiley and Sons, Inc., 1991.

[7] A. Hu and S. D. Servetto. Optimal Detection for a Distributed Transmission Array. In *Proc. IEEE Int. Symp. Inform. Theory (ISIT)*, Yokohama, Japan, 2003.

[8] A. Hu and S. D. Servetto. Algorithmic Aspects of the Time Synchronization Problem in Large-Scale Sensor Networks, 2004. ACM/Kluwer Mobile Networks and Applications. Special issue with selected papers from ACM WSNA 2003. To appear. Available from `http://cn.ece.cornell.edu/`.

[9] A. Hu and S. D. Servetto. dFSK: *Distributed* Frequency Shift Keying Modulation in Dense Sensor Networks. In *Proc. IEEE Int. Conf. Commun. (ICC)*, Paris, France, 2004.

[10] I. Maric and R. Yates. Efficient Multihop Broadcast for Wideband Systems. In *Discrete Mathematics and Theoretical Computer Science (DIMACS) series on Signal Processing for Wireless Transmission*, Piscataway, NJ, 2002.

[11] D. C. Munson, Jr., J. D. O'Brien, and W. K. Jenkins. A Tomographic Formulation of Spotlight-Mode Synthetic Aperture Radar. *Proc. IEEE*, 71(8):917–925, 1983.

[12] D. C. Munson, Jr. and R. L. Visentin. A Signal Processing View of Strip-Mapping Synthetic Aperture Radar. *IEEE Trans. Acoust. Speech Signal Proc.*, 37(12):2131–2147, 1989.

[13] A. Narula, M. D. Trott, and G. W. Wornell. Performance Limits of Coded Diversity Methods for Transmitter Antenna Arrays. *IEEE Trans. Inform. Theory*, 45(7):2418–2433, 1999.

[14] A. Scaglione and Y. W. Hong. Opportunistic Large Arrays. In *IEEE Int. Symp. Adv. Wireless Comm. (ISWC02)*, Victoria, BC, 2002.

[15] S. D. Servetto. Distributed Signal Processing Algorithms for the Sensor Broadcast Problem. In *Proc. 37th Annual Conf. Inform. Sciences Syst. (CISS)*, Baltimore, MD, 2003.

[16] P. M. Woodward. *Probability and Information Theory, with Applications to Radar*. McGraw-Hill, 1953.

Chapter 13

SENSE: A WIRELESS SENSOR NETWORK SIMULATOR

Gilbert Chen, Joel Branch, Michael Pflug, Lijuan Zhu and Boleslaw Szymanski
Department of Computer Science
Rensselaer Polytechnic Institute, Troy, NY 12180
{cheng3, brancj, pflugm, zhul4, szymansk}@cs.rpi.edu

Abstract A new network simulator, called SENSE, has been developed for simulating wireless sensor networks. The primary design goal is to address such factors as extensibility, reusability, and scalability, and to take into account the needs of different users. The recent progresses in component-based simulation, namely the component-port model and the simulation component classification, provided a sound theoretical foundation for the simulator. Practical issues, such as efficient memory usage, sensor network specific models, were also considered. Consequently, SENSE becomes an ease-of-use and efficient simulator for sensor network research.

Keywords: Wireless sensor networks, network simulation, component-based simulation.

Introduction

The emergence of wireless sensor networks created many open issues in network design [1]. The three main traditional techniques for analyzing the performance of wired and wireless networks were analytical methods, computer simulation, and physical measurement. However, many constraints imposed on sensor networks, such as energy limitation, decentralized collaboration, and fault tolerance necessitate the use of complex algorithms for sensor networks that usually defy analytical methods. Furthermore, few sensor networks have come into existence, for there are still many unresolved research, design and implementation problems, so measurements are virtually impossible. It appears that simulation is currently the primary feasible approach to the quantitative analysis of sensor networks.

ns2 (http://www.isi.edu/nsnam/ns/), perhaps the most widely used network simulator for research, has been extended to include some basic facilities to simulate sensor networks. However, one of the problems of ns2 is its object-oriented design that introduces much unnecessary interdependence between modules. Such interdependence sometimes makes the addition of new protocol models extremely difficult, which can only be mastered by those who have intimate familiarity with the simulator. The difficulties in extension are not a major problem for simulators targeted at traditional networks, for there the set of popular protocols is relatively small. For example, Ethernet is widely used for wired LAN, IEEE 802.11 for wireless LAN, TCP for reliable transmission over unreliable channels, etc. For sensor networks, however, the situation is quite different. There are no such dominant protocols or algorithms and there will unlikely be any soon. A sensor network is often tailored to a particular application with specific features, so it is unlikely that a single algorithm can always be the optimal one under various circumstances.

Many other publicly available network simulators, such as J-Sim [4], SSFNet (see for example http://www.ssfnet.org), Glomosim [14] and its commercial descendant Qualnet, attempted to address problems that were left unsolved by ns2. Among them, J-Sim developers realized the drawback of object-oriented design and tried to attack this problem by inventing a component-oriented architecture. However, they chose Java as the simulation language, inevitably sacrificing the efficiency of simulation. SSFNet and Glomosim focus on parallel simulation, with the latter tailored specifically to wireless networks. They do not appear superior to ns2 in the respects of design and extensibility.

SENSE (SEnsor Network Simulator and Emulator) aims to be an efficient and powerful sensor network simulator that is also easy to use. We identify three most critical factors in its design as *extensibility*, *reusability*, and *scalability*. We distinguish also three types of users as *high-level users*, *network builders*, and *component designers*. In the next section, we explain what each factor implies and how SENSE meets the different needs of all users. In the sections that follow, we present in details the design decisions and implementation that are centered around these design factors and that take full consideration of needs of all three types of users. Finally, we will compare the performance of SENSE with that of NS using the flooding simulation as a benchmark.

1. Design Philosophy

1.1 Extensibility, Reusability and Scalability

The enabling force behind the fully extensible network simulation architecture in SENSE is the recent progress in component-based simulation [12]. A *component-port model* frees simulation models from interdependence usually found in an object-oriented architecture, and a *simulation component classification* naturally solves the problem of handling simulated time. The component-port model makes simulation models extensible: a new component can replace an old one if they have compatible interfaces, and inheritance is not required. The simulation component classification makes simulation engines extensible: advanced users have an option of developing new simulation engines that meet their special needs.

The removal of interdependence between models also promotes reusability. A component developed for one simulation can be used in another if it satisfies the latter's requirements on the interface and semantics. In SENSE, another form of reusability is made possible by the extensive use of C++ template. A SENSE component is usually declared as a template class so that it can handle different types of data, depending on the type parameters used to instantiate the component.

Unlike many other parallel network simulators, especially SSFNet (see, for example http://www.ssfnet.org) and Glomosim [14], parallelization will be provided as an option to the users of SENSE. This decision was based on our belief that completely automated parallelization of sequential discrete event models, however tempting it may seem, is impossible. Even if it were possible, it would be doomed to be inefficient as compared to hand-tuned parallel code. Therefore, parallelizable models must require much more effort and time than sequential models, while many users are not interested in parallel simulation at all. In SENSE, a parallel simulation engine will be capable of executing an assemblage of compatible components. If a user is content with the default sequential simulation engine, then every component in the model repository can be reused.

1.2 High-Level Users, Network Builders and Components Designers

High-level users solely rely on the model repository and network template library from where they can retrieve various network models and configurations to construct a sensor network simulation. For them, the process of building a simulation merely consists of selecting appropriate models and templates and perhaps changing some parameters. Such users may not need any programming skills. Extensibility and reusability are not their concerns, but they may want the simulations to be scalable.

The network builders are not satisfied with the available network templates, but they still rely on the model repository to obtain network models. They may need to create new network topologies and traffic patterns. These users may not have immediate or knowledge of popular programming languages, such as c/c++, Java. Extensibility is not an issue for them, since they are not interested in modifying the existing models. However, models must be reusable so that they can be plugged into many simulations.

The component designer often intend to modify available models or even build new ones from scratch. For example, they can develop a proprietary MAC layer protocol which replaces the standard one. Their main concern is the extensibility; how easily existing models can be extended or replaced determines the willingness of these users to use the simulator. Reusability may or may not be an issue, depending on whether the new model is intended to be used in other simulations. The biggest challenge of the design for these users is to make the modeling process smoother, faster, and more reliable. The simulator should provide facilities to speed up checking, debugging, and verification of the models; there must be visualization tools to help identify any problems quickly; there must be standards that these users can follow in order for the models to be more accessible by others.

2. Component-Based Design

SENSE is built on top of COST [2], a general purpose discrete event simulator. The design of COST was largely influenced by the new understandings of both component-based software architecture and component-based simulation. Specifically, a component-port model was proposed to allow complex software systems to be built as a composition of components. Later, it was extended to the simulation domain where components are categorized into different types, based on how simulated time is dealt with.

2.1 Component-Port Model

In the component-port model, a component communicates with others only via *inports* and *outports*. An inport implements a certain functionality, so it is similar to a function. In contrast, an outport serves as an abstraction of a function pointer: it defines what a functionality it expects of others.

The fundamental difference between an object and a component in the component-port model is that the interactions of a component with others can be fully captured by the interface, while this is not the case for an object. For instance, an object is allowed to call member functions of any other object if it keeps a pointer or a reference to that object. Such communication, however, is not reflected in the interface or declaration of the object, and becomes manifest only when the implementation code is being examined. The resulting problem is that any function call to external objects will introduce implicit interdependence between objects, preventing the object from being reusable.

The existence of outports distinguishes components from objects. Outports impose constraints on the dynamic runtime interaction between components. The important consequence of their existence is that the development of a component can now be completely separated from the application context in which the component will be used, leading to truly reusable components. Besides, components become more extensible, because there are fewer constraints on a component that provides the desired functionality. For instance, in an object-oriented environment, if an object A is to be replaced by another object B, object B has to be derived from A. In the component-port model, this constraint is no longer necessary. Any component providing the satisfied functionality can be used, regardless of its component type.

Implementing Components. The subsequent task for us is to implement the component-port model with C++, a programming language that is usually regarded as object-oriented. Fortunately, we found template-based techniques can be utilized to archive this goal, although there are certain limitations due to the object-oriented features of the language.

First, we declare an *mfunctor* class that represents function objects for member functions of class *TypeII*. *TypeII* is the main component class, and we will explain why it is so called later in this section. The *mfunctor* class overrides the *operator()* function, so it can be called the same way as a normal function. Since it keeps a pointer to the component, it can be used to call the member function of any object derived from *TypeII*, if initialized correctly.

```
template <class T>
class mfunctor
{
 public:
  typedef void (TypeII::*funct_t)(T&);
  mfunctor(TypeII* _obj, funct_t _f)
    :obj(_obj),f(_f) {}
  void operator() (T& t) { (obj->*f)(t); }
 private:
  TypeII* obj;
  funct_t f;
};
```

The *inport* class is just a wrapper class that extends *mfunctor* so that the latter can be more conveniently initialized and invoked. To initialize an inport, a pointer to the component and a member function must be provided.

```
template <class T>
```

```
class inport
{
 public:
  void Setup(TypeII * c, mfunctor<T>::funct_t f)
  {
    functor = new mfunctor<T>(c,f);
  }
  void Write(T& t) { (*functor)(t); }
 private:
  mfunctor<T> * functor;
};
```

The *outport* class maintains a pointer to the inport to which it is connected. The *Connect()* function can be called to initialize this pointer. When the *Write()* function of *outport* is called, the *Write()* function of *inport* will be called, which in turn will invoke the member function of the component that was used to initialize the inport.

```
template <class T>
class outport
{
 public:
  void Connect(inport<T>&_in) { in=&_in;}
  void Write(T& t) { in->Write(t); }
 private:
  inport<T>* in;
};
```

One drawback of implementing components as stated above is that the inter-component communication may become quite costly, as the C++ compiler cannot completely optimize away the overhead of these function calls. However, it is possible to develop an optimization technique which can eliminate such communication overhead by merging components together so that the function to be called can be directly embedded into the code that makes the call, much the same as how inline functions work.

Another problem with the above implementation is that member functions are limited to take only one argument, as in standard C++ template classes with different numbers of template parameters cannot be given the same name. This problem can be solved by the use of wrapper classes around several arguments to make them appear as a single argument.

Components for Sensor Network Simulation. The component-port model gives the users a great deal of freedom in configuring sensor nodes. Figure 13.1

shows the internals of a typical sensor node. The sensor node is a composite component. It consists of a number of smaller primitive components, each implementing a certain functionality. Normally a sensor node has some layered network protocol components, a power component and a battery component both of which are related to power management, and others such as mobility and sensor. The inports and outports of the sensor node component are directly connected to the corresponding inports and outports of internal components.

This structure, however, is by no means the only one that users must strictly follow when they are building their own nodes. The user can freely remove or add a component, as demanded by the particular goal of the simulation. For instance, the network protocol stack can be either simplified by removing the net component, or tuned up by adding a new transport layer without affecting any other components. A queue component can be easily added between the network layer and the mac layer to prevent packets from being dropped when the mac layer is busy transmitting other packets.

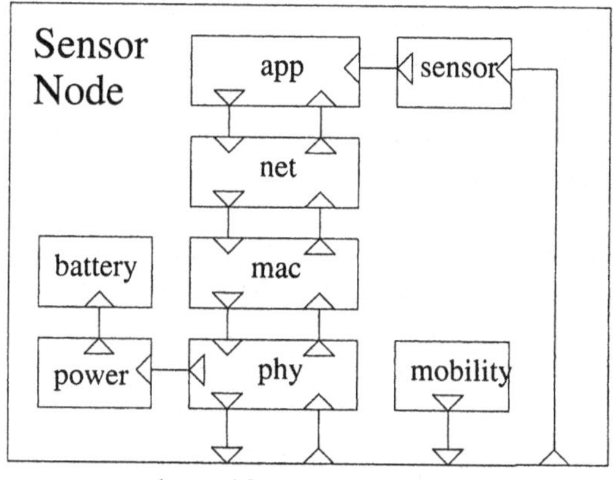

Figure 13.1. The internal structure of a typical sensor node

In theory many programming languages can be used to configure sensor nodes into a network. Configuration is as simple as setting the parameters of all components and then interconnecting their inports and outports. In this phase, components do not communicate with each other, so any object-oriented language is sufficient to perform the task. Currently, C++ is chosen to be the only configuration language, since it is also the implementation language for components. The simplicity of the configuration does allow such languages as TCL or XML to be used. In addition, it is quite natural to develop a simple scripting language specifically for the network configuration phase.

2.2 Simulation Component Classification

The component-port model clarifies the role of components in the development of general software systems. It still remains unknown, however, how the component-port model can be applied to simulation. The answer lies in a simulation component classification that naturally extends the component-port model to the simulation domain [12].

According to this classification, based on the way how simulated time is handled, simulation components are grouped into *time-independent*, *time-aware* and *autonomous* classes, also named Type I, Type II and Type III classes, respectively.

A Type I component does not have the notion of simulated time. It is passive, as it never generates events without first having received an event. A Type I component, when processing an event received from other components, may generate new events that are required to have the same timestamp as the incoming event that triggered it. Yet, the component itself is unaware of the time semantics. Neither does it know whether it is running as a part of a simulation program or a part of a non-simulation program. For this reason, a time-independent component is said to be time-unaware.

In contrast, Type II components are time-aware components. They cannot advance the simulated time themselves, but they can make a time advance request via a special object called a *timer*. Timers provide a mechanism for Type II components to generate events whose timestamp is greater than the current simulated time. To schedule such a future event, a timer is set with a time increment representing the difference between the current simulated time and the timestamp of the future event. As soon as the specified simulated time increment elapses, the component where the timer resides will be notified and then forced to process the activated event.

Type III components are named autonomous components because they maintain their own simulation clock themselves. A *clock* indicates the simulated time throughout the simulation. A sequential simulation is a Type III component by itself, which does not communicate with other Type III components. In parallel simulation, there are usually several Type III components, each mapping to a process or thread. These Type III components have to be synchronized by certain algorithms so that they can interact with each other correctly by exchanging events.

The simulation component classification leads to a hierarchical modeling process in SENSE. Because of the composability of components, a number of components can be combined into a single component. However, this kind of composition does not change the component type. If every individual component is of Type I, so will the composite component. If at least one of them is of Type II, then the composite component will also be of Type II. A *simulation engine* changes the type of the component. A simulation has to be a Type III component, so usually building a simulation involves deployment of one or several simulation engines.

This hierarchical modeling process distinguishes SENSE from many other parallel network simulators. There, the simulation engines are often built-in, and therefore users are forced to use the simulation engines provided by the simulator designers. Advanced SENSE users are given the option of building their own simulation engines, as the particular application they are investigating may call for a specific simulation algorithm (as of the time of this writing the parallelization of the simulator is still in progress).

3. Packet Management

A network simulation is composed of two types of entities: one are the static components that simulate various network devices and the other the dynamic packets that are created, transmitted, and received by components. The previous sections all dealt with only the simulation models, and we still need a good packet management scheme to effectively manipulate the packets. It turns out that this is not a trivial problem.

Our main consideration for the packet management is that it must be memory-efficient. Memory has become the most serious bottleneck that prevents large simulation programs from running on computers equipped with limited memory. Because of the extremely slow disk access speed, programs that rely on virtual memory are often an order of magnitude slower than those that can fit into the physical memory. For this reason, we decided to design a packet management scheme that consumes as little memory as possible.

This consideration makes the packet management scheme in ns2 unsuitable. In an ns2 simulation, every packet, no matter which protocol layer it belongs to, has to occupy the same amount of memory. It works well when protocol layers (other than the top one) do not create new packets, for instance, when each protocol simply appends its header to the packet and then forwards it to the lower layer. This is often not the case, however. A lower layer protocol may break a large packet into many smaller ones, as in fragmentation; it may also create new control packets, not including the original packet from the higher layer, as in handshake. In these cases, a considerable amount of memory would be wasted if we treated all packets as if they were of the same size.

Therefore, we came up with a layered packet structure, as shown in Figure 13.2. Each layer maintains its own packets, which usually consist of a header (denoted by H) and a payload field (denoted by P). The payload field contains either a pointer to, or a copy of, the packet at the intermediate upper layer. If the size of the upper layer packet is much larger than the size of a pointer, then a pointer instead of the packet itself can be kept, represented by dotted arrows; otherwise an actual copy of the packet, represented by solid arrows, will be more convenient.

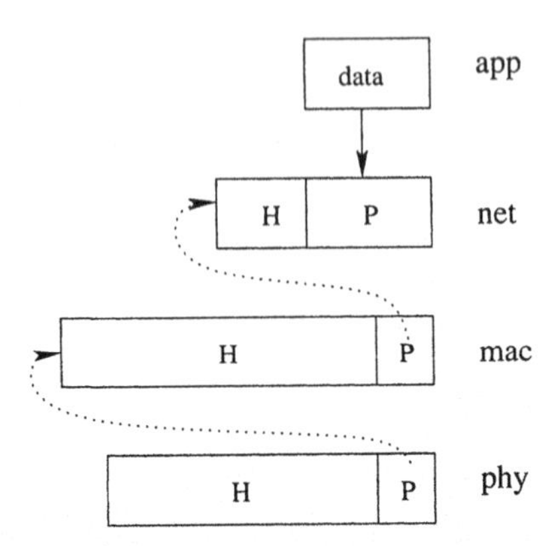

Figure 13.2. The Layered Packet Structure

Another decision we made regarding the packet management is that a packet sent by one node will be shared by all receiving nodes. This is possible because it is usually meaningless to 'modify' the receiving packet. Wireless nodes always share the communication medium with neighbors, so it is expected that one packet will often be received by many nodes. Consequently, the amount of memory saved by this approach will be considerable.

A standard programming technique, *reference counting*, is adopted to keep track of packets. When a node receives a packet, it must increment the reference count of the packet to indicate that it now partly owns the packet. When a packet is to be released, its reference count will be decremented. Only when the reference count goes to zero can the packet be actually deleted.

However, such a packet structure results in an inevitable problem. Assume a scenario in which a certain layer asks the physical layer to transmit a packet by pointer. The physical layer may successfully transmit the packet out, in which case the pointer will be forwarded to other node. However, the problem arises when the transmission fails, for instance, if there are no other nodes within the transmission range. The packet has to be destroyed by the physical layer.

This implies that the lower layer may need to be responsible for releasing the pointer to the packet sent from any higher layer, and this problem is not limited to the physical layer, since other layers may attempt to drop packets under special circumstances. In general, no reliable transmission can be guaranteed.

On the other hand, if the payload field contains not the pointer to, but a copy of the packet from the upper layer, then no operation is needed when the packet is to be dropped. For any intermediate layer, packets from the higher layer could be in the form of either pointers or plain structures. It seems that we would have to implement two components for each layer, one accepting pointers and the other copies.

Fortunately, this problem can be elegantly solved by a C++ template technique referred to as *trait*. According to Bjarne Stroustrup, a trait is "a small policy object typically used to describe aspects of a type" (for details, see http://www.research.att.com/ bs/glossary.html). In SENSE, a special packet trait class is declared which can tell if a certain template parameter is a packet structure or a packet pointer.

The declaration of this packet trait class is shown below. Basically it means that for general packets, nothing needs to be done with regard to packet deallocation.

```
template <class T>
class packet_trait
{
 public:
  static void free(const T&) {};
};
```

The *smart_packet_t* class is the main SENSE packet class defined for layers other than the top one. It consists of a header and a payload field, as well as a reference count.

```
template <class H, class P>
class smart_packet_t
{
 public:
  ...
  inline void free();
  H hdr;
  P pld;
 private:
  int refcount;
};
```

In the *free()* function of the *smart_packet_t* class, it first calls the *free()* function of the payload via the *packet_trait* class. It then decrements the reference count, and if the reference count is zero, both the header and itself will be freed.

```
template <class H, class P>
void smart_packet_t<H,P>::free()
{
  packet_trait<P>::free(pld);
  refcount--;
```

```
    if(refcount==0)
    {
      packet_trait<H>::free(hdr);
      delete this;
    }
}
```

Below is the partial specialization of *packet_trait* for pointers to *smart_packet_t*. As a result, in the *free()* function given above, if the payload contains a pointer to a smart packet, the smart packet will be freed; for all other cases nothing happens. If users are to define their own packet types and keep track of them by pointers, they should specialize the *packet_trait* class in a similar way.

```
template <class H, class P>
class packet_trait< smart_packet_t<H,P>* >
{
 public:
  typedef smart_packet_t<H,P> nonpointer_t;
  static void free(nonpointer_t* const &p)
  {
    if(p!=NULL) p->free();
  }
};
```

4. Component Repository

As the core design of SENSE has been finalized, we built an extensive set of components ranging from application layer to physical layer, as well as energy and mobility models that are specifically targeted at sensor networks.

4.1 IEEE 802.11

The IEEE 802.11 component in SENSE implemented the distributed coordination function (DCF) described in the IEEE 802.11 standard. To transmit a data packet, this MAC component first checks the size of the data packet. If the size is smaller than a predefined threshold given by a parameter named *RTSThreshold*, or if the data packet is to be broadcast, the data packet will be transmitted directly, with a proper header added. If the size is greater than *RTSThreshold*, an RTS/CTS exchange mechanism will be invoked prior to the actual data transmission, in order to reserve the medium for a period of time that is just sufficient for the entire transmission. A unicast data packet must be accompanied by an acknowledgment, but not a broadcast data packet. A transmission is deemed successful only if the acknowledgment packet has been correctly received. Each failed transmission will double the content window until it reaches the preset maximum value.

The IEEE 802.11 implementation in SENSE has the same detail level as that of ns2 (http://www.isi.edu/nsnam/ns/). However, the source code in SENSE is twice as short as that in ns2, which can be attributed to the simplicity and effectiveness of the SENSE API. For example, timers are implemented as a template class that takes the type of event as a parameter. Defining a timer in SENSE is as simple as writing a statement to instantiate the timer. On the contrary, in ns2 each timer instance needs a unique implementation of a class derived from the base timer class, which greatly degrades the efficiency and readability.

4.2 AODV

Ad-hoc on demand distance vector routing (AODV) has been well-received as a routing protocol for wireless networks. AODV's route discovery consists of setting up a forward and reverse data transmission path between two mobile nodes. After route discovery is complete, each node belonging to the established path maintains a routing table via sequenced requests and response messages. A table entry primarily consists of two IDs: one denoting the destination node and the other denoting the next-hop node along the path to the destination. The sequence numbers included in the request/response packets ensure that these routes are loop-free. Other table entry information is used to maintain route freshness, so that outdated route entries may be properly replaced. AODV's route maintenance also provides facilities for replacing damaged routes (e.g., those with broken links). Each node maintains only partial (local) route information, so full path information is never transmitted between nodes. A seminal document [8] provides more details about AODV.

The AODV implementation in SENSE is based on the most current AODV Internet draft [9]. We have implemented the operative components essential to AODV's basic operation. This set includes all steps required to actually build routes. However, selected route maintenance functions have not been included in the current simulation. For example, provisions noted in section 6.8 of [9] for handling of unidirectional links have not been implemented. This is primarily because we only assume bi-directional links in our simulation. We have not yet included full facilities for maintaining local connectivity, processing route error packets, or implementing local repair functions. All these are expected to be completed in the near future.

4.3 DSR

Dynamic Source Routing (DSR) [5] is another widely used on-demand routing protocol for wireless networks. Similar to AODV, DSR provides a mechanism of route discovery if the route from the source to the destination is unknown. But unlike AODV, after the route has been discovered, the entire route is included in the packet header, and intermediate nodes will determine the next hop by looking at the routing information contained in the packet.

An initial version of the DSR Component for SENSE has been completed which makes certain restrictive assumptions within DSR specifications. Specifically, all nodes are assumed to be bi-directional, without support for promiscuous communications, and running in a homogeneous link layer environment. Moreover, we assume that all communication links, once established, are not subject to damages, and hence error handling and route recovery are not necessary. Our testing environment currently consists of DSR running on top of the 802.11 link level component, for which all of these assumptions are valid.

As DSR matures, and new upper-level and lower-level networking components are created, a number of the current limitations will be removed. An Immediate plan is to include route error packets so that the network can recover from faulty nodes or communication obstacles. Other plans include support for the promiscuous mode operation, the optional DSR flow state extension, uni-directional links, and a data link layer which does not provide acknowledgment information for unicast packets.

4.4 Battery Models

Two battery components have been implemented in SENSE. In the *SimpleBattery* component, the discharge rate is always proportional to the power drawn from the battery, and is not dependent on the current. Its capacity is a constant defined by the simulation parameter. Let E' be the previous remaining energy and P the power consumed in the time unit, the energy remaining after a consumption period of t can be expressed as:

$$E = E' - Pt \tag{13.1}$$

In the more complex *RealBattery* component, the discharge rate becomes dependent on the current: larger current usually renders the battery discharge quicker, thus resulting in less actual capacity at the end of the usage period than the smaller current would do [7]. A discharge rate dependence parameter,

k, determines how the value of the current affects the discharge rate. More specifically, Equation 13.1 becomes:

$$E = \frac{E'}{1 + kI} - Pt \tag{13.2}$$

The *RealBattery* component also models relaxation [7], which refers to the phenomenon that a battery may gradually recover some of its lost capacity if the discharge current undergoes a sudden drop to become very small. For simplicity, we assume that relaxation only occurs if the current first sustains for a fast discharge period of at least T_R with a current larger than I_R, and then suddenly drops from above I_R to 0. Let λ be the recovery rate, g the growth ratio that can be eventually reached, then during the relaxation period the capacity is governed by the following equation:

$$E = gE'(1 - e^{-\lambda t}) \tag{13.3}$$

A restriction is imposed to ensure that the capacity after the relaxation period would not exceed the capacity right before the fast discharge period.

In this component, another parameter is provided to turn the relaxation off. If there is not relaxation, and if k, the discharge rate dependence parameter, is zero, the component regresses to the *SimpleBattery* component.

4.5 Power Model

In SENSE, the power component is responsible for power management. Currently, a *SimplePower* component has been implemented, which can operate on any of 5 modes: TRANSMIT, RECEIVE, IDLE, SLEEP, and OFF. 4 parameters specify the energy consumption rate under each of the first 4 modes, while in the OFF mode there is no energy consumption.

The power component accepts control from networking components. In response to the control signal, it can switch from one mode to another. Depending on its operating mode it also draws corresponding current from the battery.

5. Performance Comparison

To test the performance of SENSE in terms of execution speed and memory efficiency, we carried out a set of experiments that compared SENSE with ns2.

All simulations were conducted using a Dell Latitude D600 with an Intel 1.6 Ghz Pentium-M processor and 512MB 266MHz DDR SDRAM. The flooding simulation was used as the benchmark for comparison. The flooding implementation in the ns2 distribution was modified to minimize the memory usage. In the original implementation, each node maintained a hash table that stored every packet that has been received. After the modification was applied, each node would only store the latest sequence number for each source. Any packet

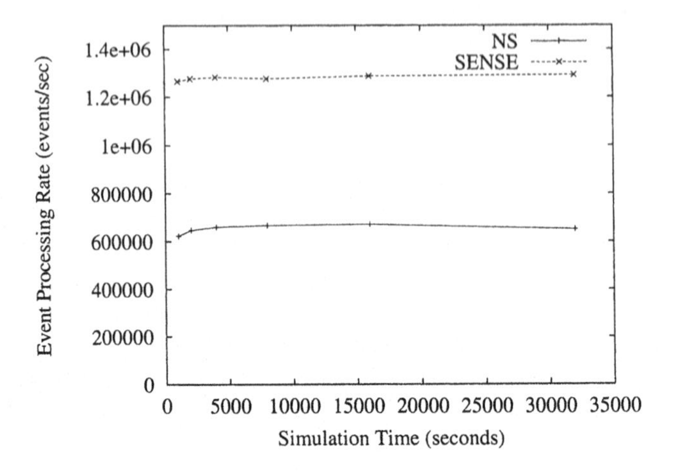

Figure 13.3. Event Processing Rate of NS and SENSE

that comes from a source with a sequence number smaller than the latest sequence number known for this source is deemed as having been received before. This modification greatly reduced memory consumption, and is in accordance with the flooding implementation in SENSE.

For the comparison, TCL and C++ scripts were written to randomly generate traffic and topology files, and both simulators were modified to read from the same input files. All nodes are running the IEEE 802.11 protocol, but using only the broadcast functionality due to the nature of flooding. Simulations were conducted to compare the two simulators execution times and memory usage under various conditions.

All NS-2 simulations were conducted using NS-2 version 2.26. A few changes were made to the flooding TCL script that comes with the ns2 distribution to disable the simulator from producing the trace file. The heap scheduler was used in both, because it is less sensitive to different time increment distributions. Unnecessary headers were also removed to minimize the size of each packet.

We compared the execution speeds of both simulators. We created a wireless sensor network containing 60 nodes, with the same random placement and a 1000m by 1000m terrain. 12 sources were randomly chosen to send packets with a length of 1000 bytes, at fixed intervals of 10 seconds. Figure 13.3 shows that SENSE is consistently twice as fast as ns2. In both simulators the number of events were roughly the same.

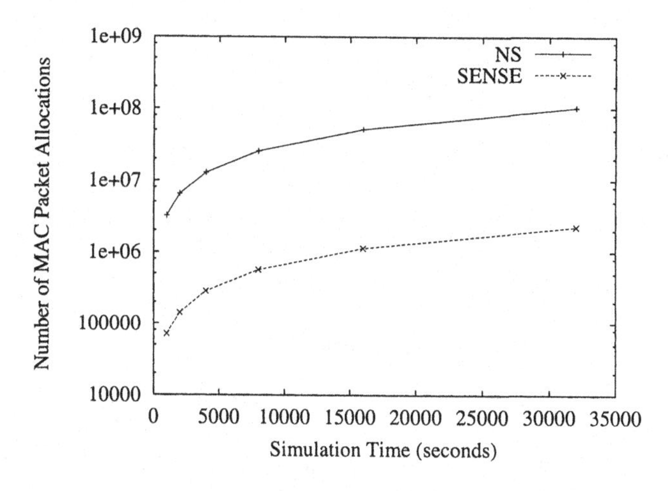

Figure 13.4. Frequency of Packet Allocation in ns2 and SENSE

The dramatic performance difference between ns2 and SENSE can be largely attributed to the ways they allocate and release packets. In ns2, when a packet is being broadcast, every neighboring node will receive a copy, so the number of packet allocations is equal to the number of received packets. In SENSE, a packet is always shared by all receivers, so the number of packet allocations is equal to the number of sent packets. In a dense wireless network, a node can usually communicate with dozens of neighbors. Consequently the number of received packets is far greater than the number of sent packets. Figure 13.4 confirms this explanation.

6. Related Work

As stated in the introduction section, the development of SENSE was largely motivated by the realization of the fundamental drawback in the object-oriented designed of ns2 (http://www.isi.edu/nsnam/ns/). Compared with ns2, SENSE is not only more efficient, as shown by last section, but also more advanced in the architecture design since SENSE greatly promotes the reusability and composability of network models.

J-Sim [4] is also claimed to be a wireless network simulation with a component-oriented architecture. However, the inter-communication efficiency was not taken as a principal design factor, and as a result the overhead is larger than in the current version of SENSE. More specifically, in every J-Sim component, a *process()* function handles incoming events for all ports, which involves dynamic dispatch of events based on the ports that they come from. However, this mechanism incurs unnecessary run-time overhead, since communication between components can be largely deduced statically from their connections.

Several other simulators devoted to wireless sensor networks have been in progress. Among them, TOSSIM [6] and Emstar [3] are similar to each other in that both are a combination of a simulator and an emulator that can facilitate the development and deployment of sensor nodes. SensorSim [10] is basically a sensor network extension of ns2, while SensorSimII [13] has been rewritten in Java but still inherited the object-oriented design. SENS [11], being developed at UIUC, is another object-oriented sensor network simulator.

7. Conclusion and Future Work

The most significant feature of SENSE is its balanced consideration of modeling methodology and simulation efficiency. In designing SENSE, we attempt to convey a belief that it is possible to build a very user-friendly simulator that is also very fast. Unlike object-oriented network simulators, SENSE is based on a novel component-oriented simulation methodology that promotes extensibility and reusability to the maximum degree. At the same time, the simulation efficiency and the issue of scalability are not overlooked. We observed that memory is the major factor that limits the size of simulation that can be actually performed, and that many other simulators contain too much overhead with respect to memory usage. The simulator is therefore memory-efficient, fast, extensible, and reusable.

SENSE is still in its active development phase. Although the core of the simulator has been gradually stabilized, it still lacks a comprehensive set of models and a wide variety of configuration templates for wireless sensor networks. Besides, a visualization tool is desirable which can quickly track down what goes wrong during the simulation. Without such a tool, the output of the simulation is hard to interpret. Visualization can also facilitate the configuration phase by allowing networks to be constructed graphically.

The problem of inefficient inter-component communication can be completely solved very soon. We have designed a component extension to the C++ language. The new language extension introduces only four keywords and four syntactic rules, with simple semantics that are easy to understand. It will not only improve the simulation speed, but also free SENSE users from the constraint that limits the number and granularity of components that can be used when efficiency is the main concern, since the inter-component communication overhead will be entirely eliminated with this new language extension.

References

[1] Akyildiz, I. F., Su, W., Sankarasubramaniam, Y., and Cayirci, E. (2002). Wireless sensor networks: A survey. *Computer Networks*, 38(4):393–422.

[2] Chen, Gilbert and Szymanski, Boleslaw K. (2002). COST: Component-oriented simulation toolkit. In *Proceedings of the 2002 Winter Simulation Conference*.

[3] Girod, L., Elson, J., Cerpa, A., Stathopoulos, T., Ramanathan, N., and Estrin, D. (2004). Emstar: a software environment for developing and deploying wireless sensor networks. In *the Proceedings of USENIX General Track 2004*.

[4] Hou, Jennifer, ying Tyan, Hung, et al. J-sim. http://www.j-sim.org/.

[5] Johnson, D., Maltz, D., and Broch, J. (2001). *Ad Hoc Networking*, chapter DSR The Dynamic Source Routing Protocol for Multihop Wireless Ad Hoc Networks, pages 139–172. Addison-Wesley.

[6] Levis, Philip, Lee, Nelson, Welsh, Matt, and Culler, David (2003). Tossim: Accurate and scalable simulation of entire tinyos applications. In *Proceedings of the First ACM Conference on Embedded Networked Sensor Systems*.

[7] Park, Sung, Savvides, Andreas, and Srivastava, Mani (2001). Battery capacity measurement and analysis using lithium coin cell battery. In *Proceedings of the 2001 international symposium on Low power electronics and design*, pages 382–387. ACM Press.

[8] Perkins, C. (1997). Ad hoc on demand distance vector (AODV) routing.

[9] Perkins, C., Belding-Royer, E., and Das, S. (2003). Rfc 3561 - ad hoc on-demand distance vector (AODV) routing.

[10] S. Park, A. Savvides and Srivastava, M. B. (2000). Sensorsim : A simulation framework for sensor networks. In *the Proceedings of MSWiM 2000*.

[11] Sundresh, Sameer, Kim, WooYoung, and Agha, Gul (2004). Sens: A sensor, environment and network simulator. In *The 37th Annual Simulation Symposium (ANSS37)*.

[12] Szymanski, Boleslaw K. and Chen, Gilbert (2002). *Lecture Notes in Computer Science, Parallel Processing and Applied Mathematics: 4th International Conference*, chapter A Component Model for Discrete Event Simulation, pages 580–594. Springer-Verlag.

[13] Ulmer, Craig. Wireless sensor probe networks - SensorSimII. http://www.craigulmer.com/research/sensorsimii/.

[14] Xiang Zeng, Rajive Bagrodia, Mario Gerla (1998). Glomosim: a library for parallel simulation of large-scale wireless networks. In *Proceedings of the 12th Workshop on Parallel and Distributed Simulations*.

Index